Electricity Generation and the Environment

The Power Generation Series

Paul Breeze—Coal-Fired Generation, ISBN 13: 9780128040065
Paul Breeze—Gas-Turbine Fired Generation, ISBN 13: 9780128040058
Paul Breeze—Solar Power Generation, ISBN 13: 9780128040041
Paul Breeze—Wind Power Generation, ISBN 13: 9780128040386
Paul Breeze—Fuel Cells, ISBN 13: 9780081010396
Paul Breeze—Energy from Waste, ISBN 13: 9780081010426
Paul Breeze—Nuclear Power, ISBN 13: 9780081010433
Paul Breeze—Electricity Generation and the Environment,
ISBN 13: 9780081010440

Electricity Generation and the Environment

Paul Breeze

ACADEMIC PRESS

An imprint of Elsevier

Academic Press is an imprint of Elsevier
125 London Wall, London EC2Y 5AS, United Kingdom
525 B Street, Suite 1800, San Diego, CA 92101-4495, United States
50 Hampshire Street, 5th Floor, Cambridge, MA 02139, United States
The Boulevard, Langford Lane, Kidlington, Oxford OX5 1GB, United Kingdom

Notices
Knowledge and best practice in this field are constantly changing. As new research and experience broaden our
understanding, changes in research methods, professional practices, or medical treatment may become
necessary.

Practitioners and researchers must always rely on their own experience and knowledge in evaluating and using
any information, methods, compounds, or experiments described herein. In using such information or methods
they should be mindful of their own safety and the safety of others, including parties for whom they have a
professional responsibility.

To the fullest extent of the law, neither the Publisher nor the authors, contributors, or editors, assume any
liability for any injury and/or damage to persons or property as a matter of products liability, negligence or
otherwise, or from any use or operation of any methods, products, instructions, or ideas contained in the
material herein.

British Library Cataloguing-in-Publication Data
A catalogue record for this book is available from the British Library

Library of Congress Cataloging-in-Publication Data
A catalog record for this book is available from the Library of Congress

ISBN: 978-0-08-101044-0

For Information on all Academic Press publications
visit our website at https://www.elsevier.com/books-and-journals

www.elsevier.com • www.bookaid.org

Publisher: Joe Hayton
Acquisition Editor: Lisa Reading
Editorial Project Manager: Mariana Kuhl
Production Project Manager: Mohana Natarajan
Cover Designer: MPS

Typeset by MPS Limited, Chennai, India

CONTENTS

CHAPTER 1

Introduction to Electricity and the Environment

The power generation industry is one of the world's largest and most extensive industrial sectors, and one of its most vital. A supply of electricity has become essential to modern life, and economic prosperity of a nation can often gauged by the level of access of its population to electrical power. The provision of that access requires investment in power stations, a transmission network to convey the energy to different regions of the nation and distribution systems to bring the electricity to the consumers. Modern renewable and distributed energy systems can provide power locally rather than from a central source but even so, security of supply still demands a high level of interconnection as well as central management and planning.

The electricity industry that supplies this resource is now highly developed. Globally there is probably around 6000 GW of generating capacity in operation, equating to thousands of power stations, and millions of kilometers of associated transmission and distribution lines. Industrial activity on this scale will inevitably have an impact on the environment at all levels at which it operates.

The earliest power stations, built at the end of the 19th century, were based on either hydropower or used oil or coal to generate heat and raise steam to drive a turbine. Both were extensions of established technologies; water had been exploited for centuries using dams to provide water for drinking and irrigation while industrial use of coal was a century old. Coal-fired power generation became the mainstay of power generation in many countries during the 20th century, and much of the industrial growth and prosperity of the advanced nations of the world was built on the back of coal combustion. Both hydropower and coal-fired power continue to be important sources of electricity. However, the environmental impact of fossil fuel combustion, of which coal is the primary component, is now seen as a threat to global security.

Electricity Generation and the Environment. DOI: http://dx.doi.org/10.1016/B978-0-08-101044-0.00001-9

Nuclear power was added to the generating mix in the middle of the 20th century and soon presented its own environmental challenges, while gas-fired power plants began to be used widely during the last 20 years of the decade. New renewable technologies started to advance from the 1970s onwards, but these did not become economically competitive until the beginning of the 21st century. These technologies are generally more benign, environmentally, than their predecessors but the scale of their deployment brings its own problems. This was becoming apparent at the end of the second decade of the 21st century when two new technologies, wind power, and solar power were advancing rapidly. Meanwhile the complex generating mix that has evolved from the combination of all these technologies needs much more sophisticated technology to manage effectively and to fully exploit the renewable potential. The challenge for the remainder of the 21st century is to provide the population of the world with electricity, while limiting the environmental impact of its production to ensure the global community remains secure and safe.

1.1 THE AWAKENING OF ENVIRONMENTAL AWARENESS

The human species has been changing the environment for thousands of years. We no longer even recognize some of those changes, for example, the clearing of forests to create the agricultural farmlands of Europe. No one now sees these fields as forests that once were. Similar changes elsewhere are more obviously detrimental to local or global conditions. Tropical rain forests grow in the poorest of soils. Clearing them, as is happening in many parts of the world, and the ground is of very little use. Not only that but also the removal of forest cover can lead to erosion and flooding as well as the loss of ground water. Most of these effects are negative.

Part of the problem is the ever-increasing size of the human population. Where native tribes could survive in the rain forests in Brazil, the encroachment of outsiders has led to their erosion. A similar pressure resulting from the volume of human demand is at work in power generation. When the demand for electricity was limited, the effect of the few power stations needed to supply that demand was small. But, as demand has risen, so has the cumulative effect of those plants on the environment. Today, that effect is of such a magnitude that it cannot be ignored.

While power generation developments have in recent years been driven by environmental demands, this was not the case when the industry was born. Take hydropower, for example, one of the earliest sources of electric power and still a clean source. In the 19th century hydropower dams were known to fail and in rare cases these failures were catastrophic. However, the effects were local and were not considered as a major hazard that required hydropower should be treated with excessive caution globally. Nor was it considered that a stable dam and reservoir could be a potential cause of concern for other reasons. Today, any new hydropower project requires an extensive environmental assessment within which every aspect of its potential impact should be evaluated.

Coal combustion was likewise considered relatively benign until around the middle of the 20th century when concern started to be raised about its effect on air quality. Consumption of coal has grown steadily since the industrial revolution. The first sign of trouble resulting from this practice was the ever-worsening pollution in some major cities. In London the word "smog" was invented at the beginning of the 20th century to describe the terrible clouds of fog and smoke that could remain for days. That city had been prone to smog since the beginning of the 19th century but it was a great smog in December 1952 that finally stimulated action, resulting in Clean Air Acts that sought finally to limit local emissions. While much of the problem was caused by the domestic use of coal, industry was also culpable. Measures to limit the industrial effects included building tall stacks to dissipate the emissions from power plants higher into the atmosphere. This was to have its own repercussions that were finally recognized later in the century.

Nuclear power arrived in the 1950s and was heralded as the great new source of cheap, clean, and limitless power. The technology was known to be dangerous if not carefully controlled, but this was an era in which such scientific and technological advances were a sign of hope for the future. By the end of the 1970s, however, the dangers were more fully recognized, and by the middle of the 1980s, construction of nuclear power plants had all but halted in the west at least. The main cause was environmental concern.

The 1980s brought recognition of another major environmental effect of fossil fuels, acid rain. Consumption of coal had continued to increase since the 1950s, but with the use of smokeless fuel in cities and tall stacks outside, problems associated with its combustion appeared to have been solved. Until, that is, it was discovered that forests in parts of northern Europe and North America were dying and lakes were becoming lifeless. During the 1980s the cause was identified—acid rain resulting from coal combustion. Tall stacks simply shifted the problem to more distant regions. More legislation, aimed at controlling the emission of acidic gases such as sulfur dioxide and nitrogen oxides, was introduced. As with smog earlier, this was a problem that had been growing for decades. Eventually the sheer scale of the industry and the volumes of pollution it was generating became too large to ignore. During the 1980s and the 1990s, too, there were critical reviews of a range of large hydropower schemes, which eventually prompted a change in the way such projects were evaluated. Even this seemingly benign technology could be the cause of major environmental problems.

Worse was to follow. By the end of the 1980s, scientists began to fear that the temperature on the surface of the earth was gradually rising. This has the potential to affect living conditions in every part of the globe. Scientists did not know whether this change was natural or man-made.

As studies continued, evidence suggested that the effect was, in part at least, man-made. The rise in temperature followed a rise in the concentration of certain gases in the atmosphere and could be traced back to the industrial revolution of the mid-18th century. Chief among these gases—now often known as greenhouse gases—was carbon dioxide. One of the main sources of extra carbon dioxide was the combustion of fossil fuels such as coal. Since then, the evidence has hardened so that the most recent report from the United Nations Intergovernmental Panel on Climate Change (UN IPCC) has concluded with a certainty of 95% that humans are the main cause of the current global warming[1].

Action to combat the causes of global warming were slow to take off, but the United Nations finally managed to persuade nations to adopt a global agreement, the Kyoto Protocol, in December 1997, which would limit carbon dioxide emissions. Action continues to evolve under UN auspices with a more far-reaching agreement adopted in

---

[1]Climate Change 2014 Synthesis Report, Intergovernmental Panel on Climate Change, 2015.

Paris in 2015. The current target is to limit global warming to no more than $2°C$ above the temperature at around 1750, but the long-term goal must be to reduce global temperatures again.

The recognition of the dangers of global warming has brought about a slow but accelerating change in the power generation industry. The most significant has been the development of new renewable generation technologies[2], particularly wind power and solar power, which produce no net addition to the amount of carbon dioxide in the atmosphere. These are advancing rapidly, and both are beginning to challenge traditional generating technologies economically as well as environmentally. At the same time, fossil fuel combustion technology is being adapted to make it cleaner. Energy efficiency is also being targeted.

These changes will continue throughout the century, so that by the end of the century the electricity supply industry may well be unrecognizable to somebody used to the configuration today. But these changes, too, will have environmental consequences. Already there are signs of what is to come with many more wind farms and large area solar power plants found both in the landscape and offshore. Such changes will inevitably lead to new problems. An industry as significant as power generation cannot avoid being in conflict, at one level or another, with the global environment.

At the same time, electricity is vital to modern living. Therefore, unless the world is going to regress technically the supply of electricity must continue and grow. On that basis, compromises must be sought and technical solutions found that do not result in irrevocable damage. These are the challenges that the power industry faces and with it the world.

1.2 THE STATE OF THE ELECTRIC POWER INDUSTRY

The electric power industry has evolved in stages, and each stage has brought its own environmental problems. Fossil fuels, which form the backbone of the generation industry, have brought the most widespread problems although nuclear power is perhaps potentially the most devastating if a plant were to fail catastrophically. Renewable generating technologies usually have more limitedocal impact but the

---

[2]Hydropower is often considered an old renewable generation technology.

scale of development required if these are to replace traditional combustion technologies is likely to bring its own difficulties.

Table 1.1 shows a breakdown of the electricity-generating industry by energy source for 1973 and 2014, based on figures collated by the International Energy Agency (IEA). During the 40 years covered by the figures the total amount of electricity generated globally has increased from 6131 to 23,816 TWh, not far short of four times the energy in 2014 compared to 1973. Similar figures for global generating capacity are not available, but the global capacity in 1950 has been put at 213 GW, while in 2010 the comparable figure was 5082 GW[3]. By 2020 this is predicted to exceed 7000 GW.

As the breakdown by energy source in Table 1.1 shows, the largest source of electricity both in 1073 and in 2014 was coal. The fossil fuel was responsible for 2348 TWh of electrical energy in 1973 and 9791 TWh in 2014, an output increase of more than four times. In 2014 coal generated 40.8% of global electricity, up from 38.3% in 1973. Natural gas provided a relatively small amount of global power, 742 TWh or 12.1%, in 1973. In 2014 its share had risen to 5144 TWh, 21.6% of the total and the second largest contributor to global power. Oil, the other fossil fuel in the table, provided 1521 TWh in 1973, 24.8% of the total and only second to coal in terms of total output. In 2014 oil-fired generation provided 1024 TWh, 4.3% of the total. Taken together, these three fuels generated 4611 TWh in 1972, 75% of all the

Table 1.1 Global Electricity Supply by Energy Source, 1973 and 2014[4]		
Energy Source	1973 (TWh)	2014 (TWh)
Coal	2348	9717
Oil	1521	1024
Hydropower	1281	3906
Natural gas	742	5144
Nuclear power	202	2525
Other	37	1500
Total	6131	23,816
Source: *International Energy Agency*.		

[3]The evolution of global generating capacity, NRG Expert, June 2015.
[4]Key World Energy Statistics 2016, International Energy Agency.

electricity used. In 2014 the total fossil fuel production was 15,885 TWh, 66.7% of the total.

Over the same period the amount of electricity generated from nuclear power plants has risen more than tenfold. In 1973 nuclear plants accounted for 202 TWh of generation (3.3%). In 2014 this had risen to 2525 TWh, 10.6%. Nuclear power stations were still relatively new in 1973, and this was the era when their construction accelerated.

The other major source of electricity in 1973 was hydropower. Hydropower plants generated 1281 TWh that year, 20.1% of the total produced. In 2014 the output from the world's hydropower plants had risen to 3906 TWh, accounting for 16.4% of the total.

The final energy source in Table 1.1, other, includes all the renewable sources except hydropower, which is listed separately. This aggregates the output from wind power, solar power, geothermal power, biomass power plants, and marine sources. In 1973 the output from all these plants was only 37 TWh but by 2014 it had risen to 1500 TWh or 6.3% of the global total. This is still tiny compared to the total global production, but it does represent an increase of 40.5 times the output 40 years earlier.

Table 1.2 takes the same figures and breaks them down by region. In this case the breakdown in primarily between the countries of the Organisation for Economic Cooperation and Development (OECD)

Table 1.2 Global Electricity Supply by Region, 1973 and 2014[5]		
Region	1973 (TWh)	2014 (TWh)
OECD	4463	10,765
Non-OECD Europe and Eurasia	1024	1739
China	178	5715
Non-OECD Americas	166	1215
Asia	159	2620
Africa	110	762
Middle East	31	1000
Total	6131	23,816
Source: *International Energy Agency.*		

[5]Key World Energy Statistics 2016, International Energy Agency.

and other regions. The IEA is made up of 29 member countries, all members of the OECD, and includes most of the countries of the European Union as well as Norway and Switzerland, the United States and Canada, Japan, Korea, Australia, New Zealand, and Turkey.

Thus, the OECD includes the most highly developed nations in the world, and its energy demand is the greatest of any of the regions shown in the table. In 1973 the OECD nations produced 4463 TWh of electricity, 72.8% of the global total. In 2014 the aggregate production was 10,765 TWh or 45.2% of the total. While production has more than doubled over the 40-year period, the growth rate is much lower than the overall global rate of growth. By contrast the growth in other regions has been much faster.

Most notable in this respect is China. In 1973 the estimated electricity production in China was 178 TWh. In 2014 it was 5715 TWh, 24.0% of the global total and 32 times larger than 40 years earlier. Growth in Asia excluding China has been rapid too, although not on quite the same scale. From 159 TWh in 1973, production in this region rose to 2620 TWh in 2014, 16 times larger and 11.0% of the global total.

Growth in non-OECD Europe and Eurasia was much slower with production rising from 1024 TWh in 1973 to 1739 TWh in 2014. The increase in output in the non-OECD Americas was much more rapid, rising from 166 TWh in 1973 to 1215 TWh in 2014. Production in the Middle East rose rapidly as well, from 31 to 1000 TWh. Growth in Africa was significant, with output rising from 110 TWh in 1973 to 762 TWh in 2014. However, this still represents a tiny proportion, 3.2%, of global output. In terms of electricity, Africa remains the least developed region of the world.

The figures from these two tables illustrate both the scale of the problem that the world faces, the sheer size of the generating industry, and also the shift in the center of gravity of electricity production as countries in Asia have advanced and required more electric power to support their industries and populations. It also highlights the problem of coal, the largest global pollutant and yet a vital source of energy for many countries.

When trying to tackle the environmental problems created by the electricity industry, particularly with respect to global warming, this shift in the global breakdown of output brings its own difficulties.

Until the 1980s there was little concern about coal combustion, but until that period most of the electricity production across the globe had taken place in the advanced industrialized nations. They had built their economies by exploiting fossil fuels, burning them with impunity. These nations are now controlling their emissions, but other nations, such as China and India, which are now building their economies wish for the same advantages that the developed nations had. Expecting these countries to reduce their reliance on coal, for example, when this is still the cheapest source of electricity, must be balanced by an understanding of the ambitions these countries have. Technology can help, and the recent reductions in the cost of wind power and solar cells have made alternatives more accessible. Even so it will require global cooperation if the environmental disaster that looms is to be avoided.

1.3 THE ENVIRONMENTAL CHALLENGES

Power generation raises a range of environmental challenges. Some of these are global in their effects some are local. Some lie in between the two extremes.

Global warming is the greatest challenge that the industry faces today. The earth's atmosphere is gradually becoming warmer, with average global temperatures rising at an accelerating rate. The effect, which can be traced back to the beginning of the industrial revolution, is with a high degree of certainty being caused by man-made emissions into the atmosphere, primarily carbon dioxide from the combustion of fossil fuels, coal, oil, and gas. Controlling the warming means reducing and eventually eliminating these emissions. However, combustion of these fuels continues to grow and so do the emissions.

Airborne pollutants of various sorts are also produced during the combustion of fossil fuels, particularly coal. These include sulfur dioxide and nitrogen oxides, the precursors of acid rain. Nitrogen oxides can also help cause smog. Small hydrocarbon particles and unburnt hydrocarbons are another hazard associated with these fuels. However, the main source of these today is likely to be from diesel engines. Coal also contains a number of toxic metals that can be released into the atmosphere. These include mercury, lead, and cadmium.

Nuclear power does not emit any carbon dioxide, but nuclear power generation carries its own risks. The most high profile is the risk of widespread contamination with radioactive material resulting from a nuclear accident, of which there have been three significant examples since the middle of the 1970s. Less publicized but no less problematical is the need to find a way to dispose safely of nuclear waste from power plants. This has been accumulating around the world since the industry began power generation in the 1950s. A viable and publicly acceptable means of disposing this waste is badly needed.

Renewable energy challenges include accommodating large numbers of wind turbines in various parts of the world and large areas of solar generating capacity. These technologies do not have any global impact similar to that of global warming—they are considered the main solution to the problem—but they do create local problems, particularly for onshore wind farms that create a visual intrusion not everybody accepts. Hydropower plants, particularly large dam and reservoir projects, can have even greater local impact. Large schemes can inundate massive areas of land, displacing people and wildlife. They may stimulate earth tremors, and some badly conceived projects have generated large quantities of methane, a potent greenhouse gas. Biomass power plants can also be controversial because their fuel is an energy crop and its production may displace food production or even result in the stripping of rain forest.

Low-scale environmental disruption accompanies the construction of any type of power plant. Effects include increases in heavy vehicle traffic while a plant is being built, additional noise during construction and in some cases during the operation of a power plant. There are heating effects, particularly from thermal power plants that usually require some form of cooling. If air-cooling is used, locally the air temperature might rise. With water-cooling, using water from a river or the sea, the water temperature at the output of the cooling water system will rise. Combustion plants produce waste residues that require disposal. Some residues can be reused, but some have to be buried or rendered harmless in other ways.

Transmission and distribution networks are extensive and in many cases visible, like wind turbines. Transmission systems usually require rights-of-way that allow them to cross large rural regions. In urban regions they may be buried underground. Distribution systems have the same requirements but are generally even more extensive than transmission systems. Both generate electromagnetic fields that are detectable close to the source. There has been a decades-long debate about whether these fields are harmful.

The hydrogen economy represents a potential future energy economy that avoids all carbon dioxide emissions. Hydrogen can be used in the same way as fossil fuels, both as fuel for vehicles, as a domestic fuel and in power stations. However, when hydrogen burns, it only produces water, so no carbon dioxide is released. Creating a hydrogen economy requires that hydrogen can be produced in vast quantities and then made widely available via pipelines and filling stations. It is also important that the hydrogen is produced cleanly. The cleanest method is via the electrolysis of water, which in turn requires large quantities of electricity. There have been tentative beginnings made to developing this economy. However, it has yet to be proved viable. Its full environmental ramifications have yet to be explored.

CHAPTER 2

The Carbon Cycle and Atmospheric Warming

The most serious environmental issue facing the world today is global warming. The warming is the result of the amplification of a natural process in the atmosphere called the Greenhouse Effect. This amplification is caused by emissions into the atmosphere from human activities and industries of gases such as carbon dioxide, which add to the natural atmospheric load of such gases. As its name implies, the natural process that underpin the Greenhouse Effect serve to maintain the ambient temperature necessary for life by trapping solar energy in the form of heat.

Energy from the sun is vital to life on earth for two reasons. In the first place, it is this solar energy, absorbed by plants, which drives the natural processes that make life possible. Secondly, the heat from the sun maintains a temperature on the earth's surface, which is within a narrow band that allows water to exist in its liquid form on most parts of the globe. Liquid water is also vital to life.

The energy from the sun that reaches the earth's atmosphere is composed primarily of radiation at wavelengths from the ultraviolet, through visible light to the infrared. Some of this energy is absorbed during passage through the atmosphere but much reaches the earth's surface and is absorbed there, warming the land and sea, and the plant and animal life found in both. As the surface of the earth warms, so some of the absorbed radiation is reradiated as infrared radiation. This radiation passes back into the atmosphere.

Some of the reradiated heat passes through the atmosphere and escapes into space. Another part is absorbed by gases in the atmosphere that are sensitive to infrared wavelengths. It is this additional absorption of radiation that would otherwise escape into space that leads to a heating of the atmosphere which helps to maintain the ambient temperature. The gases responsible, commonly known now as greenhouse gases, keep the earth around 30°C warmer than it would be without their presence.

Electricity Generation and the Environment. DOI: http://dx.doi.org/10.1016/B978-0-08-101044-0.00002-0

The most important greenhouse gas is water vapor, and this accounts for most of the Greenhouse Effect that keeps the earth and atmosphere warm. However, there are other gases that also play a role. These include methane, nitrous oxide, and carbon dioxide. The concentrations of these gases have varied over geological time, but in recent time they have remained relatively stable, that is, until around the middle of the 18th century. Since then there has been a slow but accelerating increase in the concentration of greenhouse gases and in particular carbon dioxide.

The increase in atmospheric carbon dioxide concentration leads to more reradiated heat being absorbed, and this has led to a gradual increase in global temperatures. Rising temperatures are thought to be responsible for changes in weather patterns across the globe, for rising sea levels and the melting of polar ice caps and for a range of other effects. The long-term consequences of this, if the rise in temperature is not arrested, are potentially catastrophic.

There is a range of sources for the carbon dioxide that is being released into the atmosphere. However, the two most important since the 1750s have been land use changes and the combustion of fossil fuels. In 2015 roughly 90% of all atmospheric carbon dioxide emissions came from the burning of fossil fuels. Coal is responsible for the largest contribution, followed by oil and natural gas. Much of the oil is burnt as transportation fuel, but most coal and a large part of all the natural gas is burnt each year to provide electric power. Reducing the emissions from the power sector is therefore vital if the growth in the atmospheric concentration of carbon dioxide is to be tamed.

2.1 THE CARBON CYCLE

Carbon dioxide, the key global warming gas, is also part of an important set of processes that take place on our planet, processes that are known collectively as the carbon cycle. Carbon itself is an essential component of life. It is the element that forms the skeleton of the most important molecules that enable life to exist and propagate. The carbon cycle is therefore in one sense, the cycle of life.

The carbon that is found in our universe today was created after the big bang that formed all matter and space. It can be identified in stars and planets as well as on comets and meteorites. It is a relatively abundant element, with the fourth highest abundance in the Milky Way after hydrogen, helium, and oxygen. On the earth, the element is found in all plants and animals, and there is an enormous amount in the atmosphere. More is held within the soil and the surface layer of the earth, and further vast quantities are contained in the world's oceans. Additional carbon lodged within the earth includes carbonates in rocks and large deposits of fossil fuels, the remains of plants that lived and died millions of years ago.

Although the largest part of the carbon on our planet is sequestered in the earth, a significant amount of the global carbon takes part in the carbon cycle. The cycle itself is a balance between processes occurring in plants and in animal life and equilibrium concentrations in places such as the oceans. This is shown diagrammatically in Fig. 2.1. Each part of the cycle relies on the others. Plants take carbon dioxide from the atmosphere and use it to create complex molecules, the reactions driven with energy harvested from the sun by photosynthetic

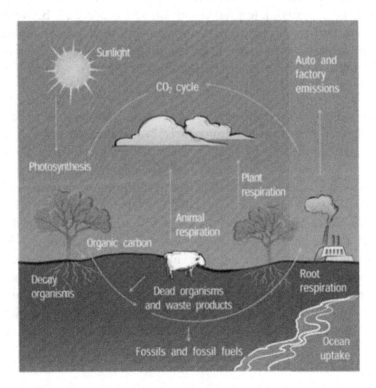

Figure 2.1 A schematic representation of the carbon cycle. US National Center for Atmospheric Research and UCAR Office of Programs.

processes. These plants then form the food for a range of microorganisms and animals, which take energy by "burning" the complex molecules made by the plants, using oxygen from the air in the process. This again releases carbon dioxide that both animals and plants produce during respiration. In this way carbon is cycled between the atmosphere and the biosphere.

To give some idea of the scale of these processes and the quantities of carbon involved, there are roughly 2200 Gt of carbon contained in vegetation, soil, and various other types of organic material on the surface of the earth. The surfaces of the world's oceans contain a further 1000 Gt, while there are 38,000 Gt in the deep oceans. There are estimated to be around 10,000 Gt held in underground deposits of coal, oil, and gas. A further 70m Gt are sequestered in rocks, often in the form of carbonates.

As already noted, the carbon cycle involves a constant exchange of carbon between the atmosphere, the biosphere, and the hydrosphere—the oceans. Approximately 60 Gt of carbon are cycled between the vegetation and the atmosphere each year and, by a process of release and readsorption, a further 100 Gt are cycled between the oceans and the atmosphere. This annual "carbon budget" is illustrated in Fig. 2.2 for the years 2006–15.

The amount of anthropomorphic carbon, carbon that is released every year by man into the atmosphere, was estimated to be 36 Gt of carbon dioxide in 2015 or roughly 9 Gt of carbon[1]. Of this, 90% was produced by combustion of fossil fuels for energy production and the manufacture of cement, as well as from the flaring of natural gas. The remaining 10% is generated by land conversion activities, primarily forest clearance and burning. Not all of this additional carbon dioxide remains in the atmosphere. Some will be used by plants; a higher carbon dioxide concentration can aid plant growth under certain circumstances. More is absorbed by the oceans. In fact, over the very long-term, the oceans would probably absorb all the excess and reestablish the long-term equilibrium. However, it would take place over the scale of hundreds of years, slowly to halt the warming now being seen.

[1]Data are from the Global Carbon Atlas, http://globalcarbonatlas.org.

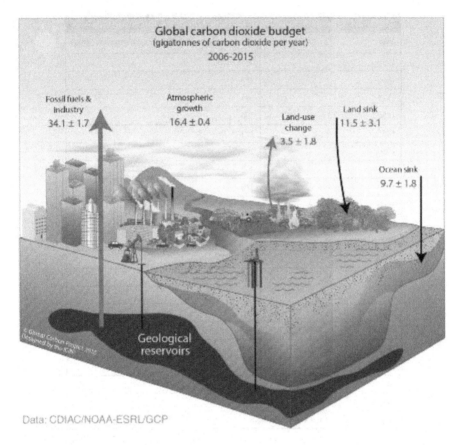

Figure 2.2 The annual carbon budget for the years 2006–15. US Department of Energy/CDIAC.

It is clear, when looking at the amounts of carbon involved, that the anthropomorphic carbon dioxide production is a small part of the whole. Nevertheless, it is sufficient to affect the overall balance, and this is enough to cause the global warming that is now being observed.

2.2 ATMOSPHERIC CARBON DIOXIDE CONCENTRATIONS

The concentration of carbon dioxide in the global atmosphere in preindustrial times was probably between 260 and 270 ppm according to measurements from such sources as polar ice cores although estimates vary. The start of the industrial era is usually placed at 1750 and the concentration then has been estimated at 277 ppm. This is shown in Table 2.1 that contains global concentrations from 1750 until 2016.

Table 2.1 Global Atmospheric Carbon Dioxide Concentrations[2]	
Year	Atmospheric carbon dioxide concentration (ppm)
1750	277
1800	283
1825	285
1850	290
1875	293
1900	297
1925	302
1950	310
1960	317
1970	326
1980	339
1990	354
2000	369
2010	390
2011	392
2012	394
2013	396
2014	399
2015	401
2016	404
Source: *Business Insights, Manua Loa*	

After 1750 the concentration began to rise slowly. It is thought that at first this was due to changes in the use of land, such as forest clearance to create croplands. Combustion of fossil fuels was known before this time, but this did not become the dominant contributing factor until around 1920.

The slow concentration rise saw the level increase by around 6 ppm between 1750 and 1800 and 7 ppm between 1800 and 1850 when the estimated mean concentration had reached 290 ppm. By 1900 the concentration had climbed by another 7–197 ppm. The concentration

[2]Figures from 1750 to 1950 are taken from the Future of Carbon Sequestration, Business Insights, 2006. From 1960 to 2016, the figures are based on June, seasonally adjusted fit figures from Manua Loa: CD Keeling, SC Piper, RB Bacastow, M Wahlen, TP Whorf, M Heimann, and HA Meijer, Exchanges of atmospheric CO_2 and CO_2 with the terrestrial biosphere and oceans from 1978 to 2000.

continued to climb slowly at the beginning of the 20th century, reaching 310 ppm in 1950, an increase of a further 20 ppm in 50 years. By then the increase in the concentration was beginning to accelerate as combustion of fossil fuels grew. In 2000 the mean concentration was 369 ppm, which is 59 ppm higher than in 1950, and by 2010 it reached 390 ppm, an increase of 31 ppm in 10 years. The mean monthly concentration actually exceeded 400 ppm for the first time in May 2013, according to figures from the Manua Loa observatory in Hawaii. The figures in Table 2.1 are for June of each year, and it was not until 2015 that the June figure breached 400 ppm. In June 2016, the concentration reached 404 ppm.

Fig. 2.3 is a complement to Table 2.1 and shows a continuous set of annual concentrations for carbon dioxide in the atmosphere from 1958 to 2016, as measured at the Manua Loa observatory. These show the continual rise, year-by-year, with an annual seasonal cycle imposed on top of the continuous increase. As a result of this and of the release of other greenhouse gases, the current atmospheric concentrations of carbon dioxide, methane, and nitrous oxide are judged to be unprecedented in the last 800,000 years, the IPCC concludes.

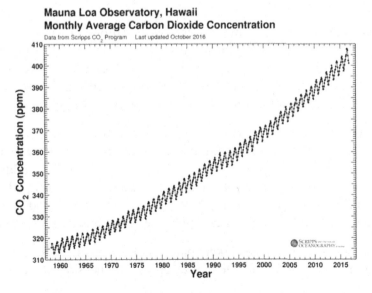

Figure 2.3 Atmospheric carbon dioxide concentrations, 1958–2016.

The link between carbon dioxide concentrations and global warming is based on a correlation between the increase in carbon dioxide and other greenhouse gas concentrations in the atmosphere and the increase in global temperatures. The latter is determined from a range of observations including seawater temperatures and air temperatures. According to the IPCC the period from 1983 until 2012 is likely to be the warmest 30-year period of the last 1400 years. Globally averaged combined land and ocean surface temperatures show a linear trend of warming of 0.85°C between 1880 and 2012. However, because there are natural variations in global temperatures, no link between the two can be claimed with absolute certainty. However, the most recent assessments suggest that the link is 95% certain[3].

2.3 THE EFFECTS OF GLOBAL WARMING

The warming of the global atmosphere can be seen from a number of measurements and observations. For example, as already noted earlier, surface temperatures have risen by 0.85°C between 1880 and 2012. The higher temperature leads to additional energy storage at the surface of the earth. More than 90% of the accumulated energy is stored in the world's oceans and only around 1% in the atmosphere itself. Most of this additional ocean energy resides in the top 75–100 m.

One result of higher atmospheric concentrations of carbon dioxide is that more is absorbed by the oceans. This has resulted in an increase in acidity. Estimates suggest that the pH of the surface layer of the oceans has increased by 0.1% or an increase of 26% in hydrogen ion concentration. Higher acidity affects many ocean-living species.

More obvious changes can be seen in weather patterns. For example, there has been increased rainfall across mid-latitude land areas in the Northern Hemisphere during the last century. In addition, Greenland and Antarctic ice sheets have been losing mass, glaciers have shrunk, and the extent of the Arctic sea ice has decreased. Between 1901 and 2010 the mean global sea level has risen by 19 cm.

The changes that have taken place so far are likely to continue, even if the emissions of greenhouse gases can be slowed or halted.

[3]Climate Change 2014 Synthesis Report, Intergovernmental Panel on Climate Change, 2015.

However, continued emissions at anything like the current level could have severe and potential catastrophic effects on global climate. Under all scenarios considered by the IPCC working groups, surface temperatures will continue to rise for the rest of this century. This is likely to lead to more frequent and extreme heat waves, while extreme rainfall is likely to become more common and more widespread. Oceans will continue to warm and become more acid, and sea levels will continue to rise.

All these changes generate stress in ecosystems. Some species will be at risk of extinction as a result. Oceans and seas are particularly at risk. Evidence from historical changes in climate suggests that smaller changes than those taking place today have led to significant ecosystem changes and species extinctions.

One of the great fears is that the changes being wrought by higher global temperatures today and through this century will continue for centuries even if greenhouse gas emissions are controlled. This could lead to some irreversible change in global conditions. While this prognosis is bleak and threatening, the global ecosystem has proved highly adaptable in the past and so there must be a strong possibility that it will prove so in the future. Whether such a world would still be a comfortable home for humans would be a matter of guesswork today. However, the scientific evidence available suggests that there is no room for complacency.

The conclusion is that whatever the correlation between anthropomorphic greenhouse gas emissions and global temperatures, human activity is leading to changes in the concentration of greenhouse gases in the environment and this change is unnatural. It should be in the interest of every individual on the planet that this change is not allowed to continue.

CHAPTER 3

Greenhouse Gas Emissions and Power Generation

As will already be clear from the preceding chapters, human activity has been adding to the amount of greenhouse gases in the atmosphere and of carbon dioxide in particular, since around 1750. Since early in the 20th century, the main contributor to this has been the combustion of fossil fuels.

Table 3.1 shows figures from 1990 until 2013 for the absolute amounts of carbon dioxide released into the atmosphere by this activity, both globally and regionally. The figures are from a report by the PBL Netherlands Environmental Assessment Agency[1] and the European Commission's Joint Research Centre.

The global figures in the second column of the table indicate that total emissions were 22.7 Gt of CO_2 in 1990. This rose to 23.6 Gt in 1995 and 25.4 Gt in 2000. By 2005 total emissions were 29.4 Gt, by 2010 they were 33.0 Gt, and by 2013 they had reached 35.3 Gt.

Table 3.1 Annual Global Carbon Dioxide Emissions (Giga tons of CO_2)[2]						
Year	Global	United States	EU	Japan	Russian Federation	China	India
1990	22.7	5.0	4.3	1.2	2.4	2.5	0.7
1995	23.6	5.3	4.1	1.3	1.8	3.5	0.9
2000	25.4	5.9	4.1	1.3	1.7	3.6	1.1
2005	29.4	5.9	4.2	1.3	1.7	5.9	1.3
2010	33.0	5.5	3.9	1.2	1.7	8.7	1.8
2013	35.3	5.3	3.7	1.4	1.8	10.3	2.1
Source: PBL Netherlands Environmental Assessment Agency/European Commission Joint Research Centre						

[1]Trends in global CO_2 emissions Report, PBL Netherlands Environmental Assessment Agency, 2014.
[2]Trends in global CO_2 emissions Report, PBL Netherlands Environmental Assessment Agency/European Commission's Joint Research Centre, 2014.

Electricity Generation and the Environment. DOI: http://dx.doi.org/10.1016/B978-0-08-101044-0.00003-2

Separate figures from the International Energy Agency show that energy-related greenhouse gas emissions of carbon dioxide have remained almost flat at 32.1 Gt from 2013 to 2015.

It is worth emphasizing that these figures are only for carbon dioxide. Other greenhouse gases such as methane and nitrous oxide are not included. In 2010, when total carbon dioxide emissions were 33.0 Gt, it was estimated that the total greenhouse gas emissions, including these other gases—with all figures converted into a carbon dioxide equivalent amount[3]—were 49 Gt CO_2 equivalent. This suggests that roughly two-thirds of "CO_2 equivalent" greenhouse gas emissions are from carbon dioxide and one-third from other gases.

Table 3.1 also contains figures for countries and regions that represent the biggest emitters of carbon dioxide. In 1990 the largest national emitter was the United States with total annual emissions of 5.0 Gt. The same year, the nations of the EU together emitted 4.3 Gt, China released 2.5 Gt, and the Russian Federation 2.4 Gt. US emissions rose from 1990 until 2005 when they reached 5.9 Gt but then fell to 5.5 Gt in 2010 and 5.3 Gt in 2013. EU emissions were also lower in 2013 than in 1990. In contrast, emissions in China rose dramatically over this period, reaching 3.6 Gt in 2005, 8.7 Gt in 2010, and 10.3 Gt in 2013. Indian emissions have also risen.

These figures show how the emissions landscape is changing. In 1990 the richer countries such as the United States, Japan, and those of the EU were the primary contributors to global carbon dioxide emissions. In 2013 the balance had shifted toward the big developing nations, particularly China and to a lesser extent India. Moreover, emissions in the EU have started to show a significant drop as strict emission control legislation over the whole region takes effect.

Energy-related emissions of carbon dioxide are slightly lower than the total emissions, as can be seen when comparing the emission figure in Table 3.1 for 2013, 35.3 Gt, with the figure for energy-related emissions for that year from the IEA, 32.1 Gt. Fig. 3.1 shows how energy-related carbon dioxide emissions have varied over the 40 years from 1975 (when they were 15.5 Gt) to 2015.

---

[3]Carbon dioxide equivalent is a way of converting the quantity of another greenhouse gas into the amount of carbon dioxide that would cause the same amount of global warming.

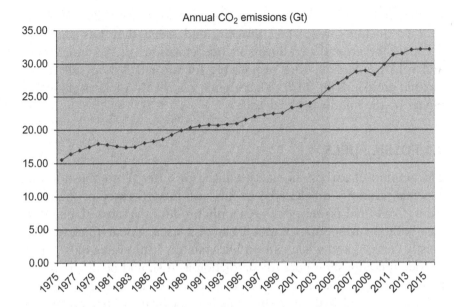

Figure 3.1 Annual energy-related carbon dioxide emissions. Source: International Energy Agency.

Table 3.2 Breakdown of Global Greenhouse Gas Emissions by Economic Sector[4,5]	
Economic Sector	**Total Greenhouse Gas Emissions (%)**
Electricity and heat production	25
Agriculture, forestry, and other land usage	24
Buildings	6
Transport	14
Industry	21
Other energy	10
Source: *IPCC*	

Table 3.2 shows a breakdown of total greenhouse gas emissions by sector for 2010 from the IPCC. The table shows that electricity and heat production accounted for 25% of the total, followed by agriculture, forestry, and other land usage. Industry emitted 21% of the total and transportation emitted a further 14%. Other sources accounted for

[4]Climate Change 2014 Synthesis Report, Intergovernmental Panel on Climate Change, 2015.
[5]The figures are based on an analysis of 2010 emissions of all greenhouse gases when the total was 49 Gt CO_2 equivalent.

the remaining 16%. Virtually all the emissions from electricity and heat production are carbon dioxide, and given that this gas alone accounts for around two-thirds of total emissions in the table, power and heat production were responsible for around 38% of all carbon dioxide emissions in 2010. That figure was likely to be significantly higher by the middle of the decade.

3.1 FOSSIL FUELS

The amount of carbon dioxide produced by different combustion fuels depends primarily on the type of fuel and the efficiency of the power plant. Coal-fired power plants generally produce the most carbon dioxide, both on a specific output basis and as a global total. Coal is mostly carbon, and when carbon burns in air, it produces only carbon dioxide if combustion goes to completion as it should be in an efficient power plant. Methane, the main component of natural gas, contains both carbon and hydrogen, and when it burns, it produces a mixture of carbon dioxide and water vapor.

It is possible to estimate the amount of greenhouse gas emissions produced by different types of power plant for each unit of electricity they produce. Life cycle emission figures for a range of the most important power generating technologies are shown in Table 3.3. These figures are in kilogram CO_2 equivalent, a figure that aggregates carbon dioxide with all the other greenhouse gases, their amounts converted into the amount of carbon dioxide that would produce the same warming effect (Life cycle studies will be discussed in Chapter 9).

Table 3.3 Life Cycle Greenhouse Gas Emissions From Key Generating Technologies[6]	
Technology	Emissions (kg CO_2 equivalent/MWh)
Coal	1001
Natural gas combined cycle	469
Utility solar photovoltaic	46
Hydropower	4
Onshore wind	12
Nuclear	16
Source: *IPCC*	

[6]Figures are from IPCC 2011 reports, published by Wikipedia at https://en.wikipedia.org/wiki/Life-cycle_greenhouse-gas_emissions_of_energy_sources.

The figures in Table 3.3 indicate that a coal-fired power plant produces the greatest emissions, an average of around 1001 kg for each megawatt of electricity it produces. The pulverized coal-fired plant is the most common and potentially the most efficient type of coal-burning power plant. The best plants can achieve up to 48% efficiency although the average is closer to 43%. Other coal-fired plants are much less efficient, so that the global average efficiency according to IPCC figures is 35%. A natural gas combined cycle power plant is the most efficient type of gas-fired plant available. Typical efficiency is 56%, and the best can achieve close to 62%. Natural gas-fired power plants, on average, produce 469 kg of CO_2 equivalent for each megawatt of power they generate.

The other technologies listed in Table 3.3 produce much less carbon dioxide. Of the renewable technologies in the table, a utility solar photovoltaic power plant produces the largest quantity of emissions over its lifetime, 46 kg/MWh. This relatively high figure is a result of the energy needed to manufacture pure silicon, which is required for solar cells. Today, this is made using electrical power and because of the global energy mix, much of that comes from power stations that produce carbon dioxide. A wind plant, with emissions of 11 kg/MWh, produces only one quarter of the emissions of the solar cell plant, while hydropower, with 4 kg CO_2 equivalent/MWh, has the least impact as far as carbon dioxide emissions are concerned.

Nuclear power also produces very low emissions of carbon dioxide. For this reason, nuclear technology is often promoted as a low carbon technology. However, there are other environmental issues associated with nuclear power, which make it less attractive to many people.

3.2 CONTROLLING CARBON DIOXIDE EMISSIONS

As figures from Table 1.1 in Chapter 1, show, fossil fuels including coal, oil, and natural gas together accounted for around two-thirds of all the electricity generated in 2014. This is indicative of a world that is massively reliant on these fuels for its electric power. This chapter has shown how these fuels are also one of the major contributors to the rise in the concentration of carbon dioxide in the atmosphere. If the rise in concentration of greenhouse gases is to be halted and eventually reversed, then power generation will have a major role to play by limiting its emissions.

There are already various initiatives intended to achieve this aim. The most important of these, globally, is the Intergovernmental Panel on Climate Change that was set up in 1988 by the World Meteorological Organization and the United Nations Environment Programme. The role of the IPCC is to provide scientific analysis of the climate and of changes occurring to the climate to provide to policy makers at all levels. In particular, the reports and studies from the IPCC are used to underline the negotiations that take place under the United Nations Framework Convention on Climate Change (UNFCC). The UNFCC came into force in 1994, and today there are 197 nations that have ratified the convention.

The UNFCC is responsible for trying to negotiate global agreements to limit greenhouse gas emissions and so control climate change. The organization holds regular meetings and has so far negotiated two major global treaties, the Kyoto Protocol and its successor, the Paris Agreement. The Kyoto Protocol was adopted in Kyoto, Japan, in December 1997 and finally entered force in February 2005 after it had been ratified by 55 nations accounting for 55% of total global greenhouse gas emissions. The protocol was aimed primarily at developed nations, which agreed to reduce their carbon dioxide emissions relative to a benchmark of emissions in 1990.

The Paris Agreement, which succeeds the Kyoto Protocol, was adopted in Paris in December 2015 and entered force in November 2016 after it had been ratified by 116 nations. The new agreement is stronger, in principle, than its predecessor because it requires all nations to put forward plans to reduce their emissions.

3.3 CARBON TRADING AND CARBON LIMITS

While the UNFCC agreements are global, they have to be enacted nationally or regionally and that comes down to national or to regional government. There is a range of measures that can be employed at this level to try and control emissions.

One of the largest schemes in place is the EU Emissions Trading System that covers all nations in the EU. The EU approach is a "Cap and Trade" scheme. A cap is set on the total amount of carbon dioxide that can be released across the EU, and then allocations are made for individual countries and for individual power stations or industrial plants within those countries. Each power plant must then buy certificates equivalent to the amount of carbon dioxide they are permitted to release during a year. For each ton of carbon dioxide released, the plant must redeem the equivalent number of certificates. However, if the plant does not need all its certificates, it can sell them on the Emissions Trading System. Other plants that wish to emit more than their allowance can buy certificates that allow them to increase their emissions.

The cap is reduced every year, so that emissions across the region fall from year to year. However, the success of this type of system depends on the cost of transgressing. If companies do not have to pay sufficiently heavily for emitting more carbon dioxide than they have certificates for, then the system will fail. The EU system has also suffered because the cost at which certificates have traded in recent years has been too low to encourage the cutting of emissions below the cap.

There are other ways of trying to limit emissions. One is a simple carbon tax, imposed on all emissions of carbon dioxide by all plants and industries. This does not seek to limit emissions at a specific level, but it does make all polluters pay. A carbon tax can be levied on emissions directly, or it can be levied on any fuel that contains carbon, which when burned will release carbon dioxide. Carbon taxes of different sorts have been levied in several countries.

These regulatory and legislative measures can provide a framework for controlling carbon dioxide emissions. However, the actual reduction of emissions depends on changes to the way the power industry operates. These are technology-based.

3.4 TECHNOLOGICAL SOLUTIONS

If carbon dioxide emissions from power stations are to be controlled, then ways have to be found to reduce the emissions from these plants. There are two fundamentally different ways of tackling this problem. The first is to stop burning fossil fuels entirely, replacing the generating capacity with alternatives that can produce electricity more cleanly. The second is to develop ways of burning fossil fuels without emitting carbon dioxide. Both are possible but while the first is well advanced, the second is taking a prodigiously long time to develop.

The replacement of fossil fuel—based power plants with different types of power plant began to gather momentum at the beginning of the 21st century and since then has advanced rapidly. The key to this has been the development of low-cost renewable energy sources. The most successful of these so far have been wind energy and solar photovoltaic energy from solar cells. Hydropower also has a role to play, but many of the best sites for hydropower have already been exploited in the developed nations, so these have to turn to new renewable technologies. Elsewhere there is still scope for additional hydropower development.

The advance of the main new renewable technologies has been much swifter than expected, and this is thought to be one of the main factors responsible for the leveling of carbon dioxide emissions from 2013 to 2015. If so, then this is a cause for cautious optimism. According to the IEA, in 2015 electricity generated from new renewable energy accounted for around 90% of all the electricity from new power plants. Of this, wind power accounted for more than half of new generation.

Meanwhile, the other approach to emission control has been slow to advance. This relies in finding ways of removing the carbon or capturing carbon dioxide during fossil fuel combustion. Three key approaches to this have been developed. The first and the simplest is to add a carbon dioxide capture system to the exhaust of a fossil fuel power plant. This scheme, known as postcombustion capture, processes the flue gases from the power station, removing all (or most) of the carbon dioxide before releasing the cleaned combustion gases into the atmosphere. So far there have been only two such projects developed, both relatively small compared to a standard coal-fired power plant.

The second way of attacking the problem of providing clean power from fossil fuel is a process called precombustion capture. This process involves converting the fuel into a hydrogen-rich gas and at the same time removing all carbon dioxide produced during the gasification process. No full-scale power plant of this type has been built. A third approach is to adopt a system called oxy-fuel combustion; instead of burning the fossil fuel in air, it is burnt in pure oxygen. The advantage of this is that the exhaust gases are almost pure carbon dioxide that is relatively easy to capture. No large-scale power plant of this type has been built either.

After carbon dioxide has been captured using one of these techniques, a way must also be found for disposing it. This is likely to be by pumping it into underground stores where it can remain forever, a process known as carbon sequestration. A few test facilities that are sequestering carbon dioxide are using exhausted oil and gas wells for the purpose, but there are a range of other underground geological features that are also suitable.

An interim fossil-fuel approach, although not a solution, has been to displace coal-burning power stations with natural gas-fired facilities. These still emit carbon dioxide, but they produced less for each unit of electricity they produce. This approach, which allows nations to reduce their emissions, has been used in several developed countries.

With the major developing nations, as well as some developed countries, still relying heavily on coal and natural gas for their electric power, it has seemed for the past two decades that carbon dioxide capture and its sequestering underground was essential if control of the emissions was to be achieved. However, the slow rate of development of the technologies for carbon dioxide capture, together with the rapid advance of renewable technologies, may mean that carbon capture technologies will now only play a small role in the future power generation mix.

CHAPTER 4

Combustion Plant Emissions: Sulfur Dioxide, Nitrogen Oxides, and Acid Rain

Carbon dioxide, the subject of the previous two chapters, is harmful to the environment as a consequence of its ability to cause global warming, but it is not a pollutant in the normal understanding of the term because carbon dioxide is naturally present in the atmosphere, independently of fossil fuel combustion. However, fossil fuel combustion can produce a range of genuinely harmful pollutants, and in most cases these will be released into the atmosphere unless action is taken to control their emission. Some of these pollutants are the result of impurities contained within the fuel. Others are a consequence of the combustion process itself.

The combustion of coal is the dirtiest of the large-scale methods of generating electricity, primarily because of the range of potential pollutants that are found within the fuel. While some high-quality coals are relatively pure carbon, many are far from pure. Impurities commonly found in coal include sulfur, bound nitrogen, volatile organic compounds (VOCs), heavy metals including cadmium and mercury, and a range of inert refractory materials. All of these can be released into the atmosphere during coal combustion if measures are not taken to remove them from flue gases.

Natural gas presents its own set of problems. The gas as it emerges from the ground generally contains between 70% and 90% methane. The rest is a mixture of higher hydrocarbons, nitrogen, water vapor, carbon dioxide, and hydrogen sulfide. Processing removes most of the impurities, and the gas that enters a natural gas pipeline usually contains more than 90% methane. The first problem with natural gas is that methane is itself a potent greenhouse gas—more effective than carbon dioxide—so any gas that is accidentally released during oil and gas extraction and transportation adds to the atmospheric load of these gases. The US Environmental Protection Agency has estimated that 3.8% of all greenhouse gas emissions in the United States between

Electricity Generation and the Environment. DOI: http://dx.doi.org/10.1016/B978-0-08-101044-0.00004-4

1990 and 2009 came from these sources. The second problem with natural gas is that its combustion generates nitrogen oxides, and these can cause a range of health and environmental problems.

Oil is used less today for power generation than either coal or natural gas, and its impact lies somewhere between the two. Heavy oil may be as dirty as coal when burnt, but lighter fuel oils are relatively clean. Their main pollutant emissions are likely to be nitrogen oxides[1], similar to natural gas. Fuel burnt in diesel engines also produces a range of damaging airborne particles, usually referred to as particulates, and these has been identified as a major problem in urban areas with large populations of diesel-powered vehicles. It is less of a problem for power generation.

4.1 AIR QUALITY STANDARDS

The problems created by these various emissions were recognized gradually during the 20th century. As they were identified, legislation was introduced in many parts of the world to limit their release. These regulations are generally linked to air quality standards. There are differing types of standards depending on what is being measured and where. Long-term air quality standards will define what is considered as the upper level of a particular pollutant concentration that is acceptable. However, a power plant exhaust will usually be permitted to contain a much higher concentration on the understanding that this will be diluted over time and distance.

Air quality regulations and emission limits for pollutants from power stations vary from region to region, but most countries enforce some limits today. These tend to be the strictest in the most developed countries such as those in Europe, Japan, and the United States. Table 4.1 contains figures for the concentrations of the various power plant airborne pollutants that are considered permissible in the EU, and in the United States, if good air quality is to be maintained. EU regulations are generally the stricter; for example, the EU expects

---

[1]Nitrogen oxides (NO_x) is the term used to describe a mixture of oxides of nitrogen that can be formed during combustion. The main oxides are nitrogen oxide, NO; nitrous oxide, N_2O; and nitrogen dioxide, NO_2. During fossil fuel combustion in power plants the main oxide produced is nitrogen oxide.

Table 4.1 Air Quality Standards		
Pollutant	EU standard/averaging period	US standard/averaging period
Sulfur dioxide	125 µg/m^3/24 h	365 µg/m^3/24 h
Nitrogen oxides	40 µg/m^3/1 year	100 µg/m^3/ 1 year
Particulate matter (PM10)	40 µg/m^3/1 year	150 µg/m^3/24 h
Carbon monoxide	10 mg/m^3/8 h	10 mg/m^3/8 h
Ozone	120 µg/m^3/8 h	150 µg/m^3/8 h
Lead	0.5 µg/m^3/1 year	0.15 µg/m^3/3 months, rolling
Cadmium	5 ng/m^3/1 year	–
Source: *EU Commission, US Environmental Protection Agency.*		

sulfur dioxide concentrations over a 24-h period to be below 125 µg/m^3. In the United States, the same standard is 365 µg/m^3. However, internationally, standards are tending to converge as the effects of even low levels of pollution on human health become more widely recognized. The PM10 particulate matter standard is for dust particles greater than 10 µm in diameter, and this is generally the standard of importance when considering dust from coal-fired power plants. There are other standards including PM2.5 for particles of up to 2.5 µm in diameter. For the EU the PM2.5 standard is 25 µg/m^3 averaged over 1 year. Table 4.1 also shows heavy metal limits. In the EU the limit for atmospheric lead concentration is 0.5 µg/m^3 averaged over 1 year and for cadmium it is 5 ng/m^3. In the United States the limit for lead is 0.15 µg/m^3 on a 3-month rolling basis. There are proposals to introduce limits on mercury emissions in the United States although these have not yet taken effect.

The figures in Table 4.1 apply to the air quality that people will encounter in the street or in their houses of offices when carrying out their daily lives. The actual emissions permitted by power plants are generally much higher than this. A power plant represents a concentrated source of pollutants, but these are released in hot gases from a tall stack, so that they should rise high into the atmosphere and become diluted before humans or other life forms come into contact with them. However, the behavior of the pollutants once they enter the atmosphere is not always predictable. The behavior of the plume of exhaust gases from a power plant stack will depend on atmospheric conditions, so that sometimes the pollutants will fall close around the plant, and at other times they may be carried across continents.

Table 4.2 EU Emission Limits for Large Power Plants	
Sulfur dioxide emissions for plants built after 2003	200 mg/m^3
Sulfur dioxide emission limits after 2016	150 mg/m^3
Nitrogen oxide emissions for plants built after 2003	200 mg/m^3
Nitrogen oxide limits after 2016	150 mg/m^3
Dust emission limits after 2016	20 mg/m^3
Proposed mercury emission limit	30 μg/m^3
Source: *EU Commission.*	

Table 4.2 shows some of the emission levels permitted within the EU for power plant flue gases. The figures are for large plants with a thermal capacity in excess of 300 MWth. The limits are less strict for some smaller plants. For sulfur dioxide the limit for plants built after 2003 is 200 mg/m^3, falling to 150 mg/m^3 after 2016. Permitted emission levels for nitrogen oxides are the same. Dust emissions are to be below 20 mg/m^3 after 2016, and there is a proposed emission limit for mercury of 30 μg/m^3.

4.2 THE FUELS

The three fossil fuels that are commonly used in combustion plants for power generation are coal, natural gas, and oil. All three fuels generate atmospheric pollutants, and coal combustion produces solid residues as well. The major pollutants can all be removed from the exhaust gases of a combustion power plant, but the capture process may itself result in further solid residues.

4.2.1 Coal

Most coals contain some sulfur. Often it is more than 3% of the coal, and it may reach as much as 10%. When the coal is burnt, this sulfur is converted into sulfur dioxide that is carried off by the flue gases. If released into the atmosphere, it can be converted into an acid. There is organic nitrogen within coal too, a residue of the organic materials from which the coal was formed. During combustion, this is converted into nitrogen oxides of various sorts including NO, NO_2, and N_2O. Another important source of gaseous nitrogen compounds in flue gases is the nitrogen in air, which can become oxidized at the high temperatures encountered within coal furnaces. Both nitrogen oxides and sulfur dioxide can be potent pollutants.

Coal usually contains a significant amount of mineral impurity too. Some of this may melt and fuse with other similar material during the high temperature combustion in a pulverized coal plant, creating a solid residue that is left behind in the combustion chamber as slag. This is eventually removed from the bottom of the furnace. However, depending upon the exact combustion conditions, a large proportion of the inert solid material may remain in small enough particles to be entrained and carried away with the flue gases exiting the boiler. These particles may contain heavy metals such as cadmium and mercury, which, if allowed to escape, will be released into the environment, so they too must be contained.

Some coals, particularly the bituminous varieties, contain large amounts of VOCs and these, or fragments of them generated by their incomplete combustion, can be released. Incomplete combustion of the carbon in coal may also lead to significant levels of carbon monoxide within the flue gases. Both carbon monoxide and organic fragments can cause environmental degradation as well as affecting human health if allowed to escape.

4.2.2 Natural Gas

The natural gas as it emerges from the ground is a mixture of hydrocarbons, carbon dioxide, nitrogen, water vapor, and hydrogen sulfide. This is processed in stages to remove the hydrogen sulfide, higher hydrocarbons, carbon dioxide, and water vapor. The resulting pipeline gas, the gas supplied to power plants, is primarily composed of methane, together with a small amount of ethane and traces of nitrogen and carbon dioxide. Methane itself is an important greenhouse gas, so any methane that escapes during extraction, processing, or transportation will add to the greenhouse gas burden in the atmosphere. Methane is a much more effective greenhouse gas than carbon dioxide. It has been estimated to be 84 times more potent during the first two decades after its release. However, it does not persist for so long. Even so, it has been estimated that around 25% of the man-made global warming experienced today is caused by methane[2]. The other main danger from natural gas combustion is the generation of oxides of

---

[2]These figures are from the Environmental Defense Fund website: https://www.edf.org/methane-other-important-greenhouse-gas.

nitrogen. These are formed from oxygen and nitrogen in air at the high combustion temperature when natural gas is burned.

4.2.3 Oil

Oil was a popular fuel for power generation during the middle of the 20th century but is much less frequently used now. Oil as it is pumped from the earth contains a range of different hydrocarbon compounds, and the composition varies from region to region. Refining of oil separates the various components into fractions according to their boiling points to produce a range of lighter and heavier oils. Important hydrocarbon fractions, ordered by increasing boiling point, are refinery gas that is used as bottled gas and consists mostly of propane or butane, petrol, naphtha, kerosene, diesel oil, lubricating oil, fuel oil, greases/ waxes, and bitumen. Petrol and diesel are used in large piston engines for power generation, and liquid fuels may also be burnt in gas turbines. Potential pollutants from these include carbon monoxide, particulates, VOCs, and nitrogen oxides. All are produced by the combustion process, and these fuels contain few chemical pollutants. Fuel oil (often called residual fuel oil or heavy fuel oil) is the material left behind after distillation of lighter fractions has been completed. It is a much dirtier fuel and may contain up to 5% sulfur and heavy metals, as well as hydrocarbons. It is used in some large, slow speed diesel engines. Combustion of this fuel can produce sulfur dioxide, nitrogen oxides, VOCs, carbon monoxide, and particulates.

4.3 ACID RAIN

Acid rain is the name given to any precipitation from the atmosphere that is abnormally acidic in nature. The term became common currency during the 1980s when the adverse effects of acidic precipitation on the environment first caused major concern. At its worst, acid rain can affect fish populations, cause degradation and even death of lakes and streams; it causes erosion on buildings and is hazardous to human health. Recognition of this eventually led to the control of the emissions responsible for the acidification.

The effect known as acid rain was first identified in Sweden in 1872, but it was not until the 1970s, again in Scandinavia, that the modern effects were clearly identified. Since then it has been studied in many parts of the world and its impact quantified. The acidity of water is classified according to its pH, a logarithmic scale of the hydrogen ion concentration. Neutral water has a pH of 7; below that is considered acid and above is alkaline. Normal rain absorbs carbon dioxide from the atmosphere, forming carbonic acid that is slightly acidic, so that the normal pH of rain is 5.6. Acid rain has a typical pH of between 4.2 and 4.4.

Higher acidification of rain is caused by the presence of sulfur dioxide or nitrogen oxides in the atmosphere. When these are absorbed by moisture in the atmosphere, they can form sulfuric acid or nitric acid, much more potent acids that carbonic acid, and these can lead to much higher levels of acidity. The effect is illustrated diagrammatically in Fig. 4.1. The level of acidity is critical for some species. For example, below a pH of 5, fish eggs cannot hatch while for frogs the critical pH is 4. Elsewhere acidic precipitation leads to dying of trees.

There are a number of natural sources of acid rain. These include sulfur dioxide released from volcanic eruptions and nitrogen oxides generated by lightning. However, the main sources of the acidic gases in the atmosphere are from human activity. The electric power industry probably makes the biggest contribution, accounting for two-thirds

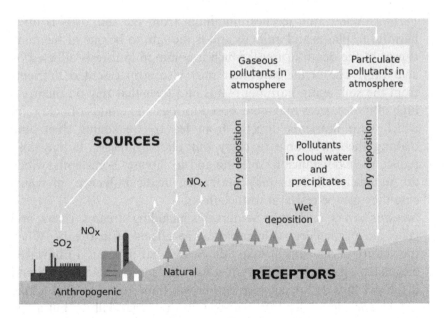

Figure 4.1 Acid rain. Modified version of Wikipedia image.

of sulfur dioxide in the atmosphere and around one quarter of the nitrogen oxides.

Acid rain can arrive at the surface of the earth as either wet or dry deposition. Wet deposition, in the form of rain, snow, fog, and hail, is what the term acid rain normally brings to mind. However, it can also fall as acidic gases and particles that are deposited directly from the atmosphere. These particles can be harmful to human health if inhaled. Dry acidic deposition will be washed off during the next rainfall, draining into the soil, rivers, and streams where it will cause damage in the same way as wet deposition.

The flue gases from power plants, carrying any acidic gases, are normally swept high into the atmosphere. Winds can then carry the gases over long distances before they are deposited again. In consequence, acid rain can be transported from nation to nation and region to region. Acidification of lakes in Scandinavia, for example, has been linked to emissions of sulfur dioxide in the United Kingdom. In the United States trans-state acid rain transportation has been common. Long distance transport makes acid rain a global problem.

Acid rain has been linked to a number of specific problems:

Forests: Acidic rain leeches aluminum from the soil. This is often harmful to plants and animals and is thought to be one of the main causes of the death of trees, which is common in forests affected by acid rain. The acidic rainwater may also strip essential nutrients from the soil, again harming plants and trees that rely on them. At high altitudes, trees may be exposed to fogs and clouds of acid rain, which strip nutrients directly from leaves, weakening them and making them less able to carry out photosynthesis. It may also reduce their resistance to freezing and to disease. Some soils, which are naturally alkaline, may buffer the acidic influence, in which case they can be resistant to the effects.

Streams, rivers, and lakes: Waterways including streams, rivers, and lakes are particularly vulnerable to the effects of acid rain. When the acidity of a steam or lake falls below that of natural rainwater, 5.6, many species struggle. A few fish can survive below pH 5 and if the pH falls as low as 4, most life will have been lost and a lake will be considered dead. In between, most species will die but a few acid-tolerant species may survive. However, this is likely to skew the ecology of the system dramatically. As with soils, not all lakes and waterways are vulnerable. Those on alkaline soils and rocks can buffer the acidity and resist its effects.

Plants and crops: Crops and plants will be affected in exactly the same way as trees and forests if acid rain removes nutrients for soils or leeches aluminum, which can damage their growth. Typical effects include stunted growth.

Buildings and man-made structures: Acid rain can both cause and accelerate the corrosion and destruction of many building materials including stone and metal. Much of the erosion and damage to city structures during the 20th century may be linked to acid rain. Some of the materials used to construct older or ancient buildings have proved to be the most vulnerable, and major monuments such as the Taj Mahal in India, Cologne Cathedral in Germany, the Colosseum in Italy, and Westminster Abbey in the United Kingdom have all suffered material damage as a result of acid rain.

Human health: Acid rain can be injurious to human health if inhaled directly. The most serious widespread problem is from acidic fog that can cause respiratory problems if inhaled. People with asthma or those who are already weak are particularly vulnerable.

4.4 CONTROLLING ACID RAIN

Acid rain can be controlled by reducing the quantities of the pollutants, sulfur dioxide, and nitrogen oxides, which enter the atmosphere as a result of industrial activity. For the power industry, the only major source of sulfur dioxide is from coal-fired power stations. Since the 1980s, a variety of systems has been developed to reduce these emissions. One approach is to clean coal before it is burned. This can remove any sulfur-bearing rocks mixed with the coal but it cannot control the sulfur actually within the coal. Another option is to burn only low sulfur coal in power stations. However, this is of limited appeal if the low sulfur coal has to be imported into a region or country; the attraction of coal-burning is that the fuel is cheap and importing can raise the price significantly.

More effective has been to treat the flue gases of a power plant after combustion has taken place. While a number of technologies have been developed since the 1980s, one of the earliest remains the most effective and the most widely used, wet flue gas scrubbing. This involves passing the flue gas up through a tower, while a slurry of water containing a material that will react with and capture the sulfur dioxide is sprayed into its path. In a well-designed wet scrubber, it is possible to capture up to 98% of the sulfur contained in the flue gases. The reagent most often used is limestone, calcium carbonate, which reacts with sulfur dioxide to form gypsum. Provided this is pure, the gypsum can be recycled as a building material.

Controlling emissions of nitrogen oxides from power plants usually involves a combination of processes. Nitrogen oxides generated during combustion can be formed from two sources, depending upon the fossil fuel. Coal contains a number of nitrogen-containing compounds, and during combustion these are converted into nitrogen oxides. This is the source of around 75% of all the nitrogen oxides generated in coal-fired power plants. The other source is nitrogen in air, which can be oxidized at the high combustion temperatures in power plants, reacting with oxygen in air to form oxides. This latter source, common to coal, oil, and gas plants, is normally tackled first.

The oxidation of air requires a high temperature, and so operating at as low a combustion temperature as possible can help alleviate the problem of oxide production. However, thermal power generation relies on heat engines that operate more efficiently, the hotter their working fluids[3] so lowering the operating temperature can affect efficiency and, therefore, economics. An alternative strategy is to ensure that within the hottest combustion zone of a furnace or burner, where nitrogen oxide generation is likely to be fastest, the amount of oxygen available is limited: in chemical terms this equates to maintaining a reducing environment in the hottest part of the combustion zone. Under these conditions the oxygen will preferentially react with the fuel instead of nitrogen. Combustion systems designed around this principle are usually called low NO_x burners. They are used in both coal and gas-fired power plants.

Low NO_x burner technology is applied to most large power plants today, but it is still not sufficient to keep the level of emissions below statutory levels, so additional measures are required. The usual approach is to add a reagent to the flue gases that will react with the

[3]The working fluid depends on the type of heat engine; for a plant that uses a steam turbine, the working fluid is steam. For a gas turbine, the working fluid is air.

nitrogen oxides, converting them back into nitrogen. The most widely used scheme is called selective catalytic reduction (SCR). SCR involves mixing a reagent, usually either ammonia or a urea derivative, with the flue gases and then passing the mixture over a metal catalyst that facilitates the reaction between the reagent and nitrogen oxides to produce nitrogen and water vapor. SCR is most often carried out when the flue gas has cooled between 250°C and 430°C, but if the reagent is introduced into the flue gas stream when it is still very hot, then the reaction will occur spontaneously and without the need for a catalyst, although with lower efficiency than SCR can achieve. Depending upon the concentration of nitrogen oxides in the flue gas, SCR can remove up to 90% of the nitrogen oxides present.

4.5 FOSSIL FUEL PLANTS AND SMOG

Thick, often acrid clouds of polluting mist and fog associated with fossil fuel combustion have been known for at least two centuries, possibly longer. These clouds are damaging if inhaled, a situation impossible to avoid while one is present. For the most vulnerable, they are deadly.

Historically, these poisonous clouds were caused primarily by the industrial and domestic combustion of coal and were usually found in urban areas where the smoke and sulfur dioxide combined with mist to create a thick blanket. This was christened smog (smoke-and-fog) in the early 20th century. Clean air legislation and controls on the emissions from coal-burning industrial and power plants have reduced the incidence of this type of smog in many parts of the world, although countries such as China and India still experience smog in major cities.

Another type of smog, photochemical smog, is also created from pollutants generated during fossil fuel combustion. This is a more modern type of pollution. In this case one of the main causes is nitrogen oxides and especially nitrogen dioxide. This particular chemical can react under the influence of sunlight to create ozone that is harmful in itself and can react with other hydrocarbon pollutants to create a range of chemicals that are injurious to human health. VOCs are also implicated in smog production and cause further health issues.

The main source of the precursors of photochemical smog is emissions from vehicles, but power plants also produce the same types of pollutant. A typical photochemical smog is a thin, brown haze.

4.6 DUST AND PARTICULATES

Small solid particles—particulate material that is often known simply as particulates—is produced when most fossil fuels are burned. The amount and type of the particulates depends on the fuel and plant. Coal-fired power stations can release large quantities of fly ash, a residue of the combustion process that is carried away with the flue gases. This is composed of incombustible material, and the particles may also contain toxic metals. It can be visible as a plume of smoke and may fall as a fine dust on the surrounding area. Smaller particles can be carried much further. Natural gas and oil-fired power plants may also generate particulate material, but its composition is likely to be different, based primarily on unburnt carbon or organic material.

The particles that are produced during combustion are graded according to their size. PM_{10} refers to particles with a size of $10\,\mu m$ or less. $PM_{2.5}$, likewise refers to particles smaller than $2.5\,\mu m$. It is normally considered that particles larger than $10\,\mu m$ can be filtered from air passages when inhaled, while smaller particles are capable of penetrating deeper into the lungs. The smaller the particles, the more dangerous they are considered, with those under $1\,\mu m$ (PM_1) and lower, particularly damaging.

Coal plants are an obvious and visible source of particulate material, and virtually all modern power stations now include systems to remove dust from their flue gases (unfortunately some smaller and older combustion plants do not and these still present a major environmental hazard). There are two main ways of removing the dust from coal plant flue gases. The first is to use a mechanical filter called a baghouse. This is simply a fabric filter through which the flue gas passes. Dust particles are removed while the gases pass on. Devices of this type can be more than 99% efficient in removal of the particulates.

The alternative is to use a device called an electrostatic precipitator (ESP). This is a special chamber with charged wires and plates that allow a large electrostatic voltage to be applied to flue gas as it passes through the chamber. The voltage induces a charge on the dust particles, and they are then collected on the plates in the ESP from where the dust is removed by "rapping" them to make the collected dust fall into the bottom of the chamber. ESPs are capable of removing up to 99.7% of the dust in the combustion gases from a coal-fired power station. However, even with these facilities, coal-fired power plants can still release significant quantities of both PM_{10} and $PM_{2.5}$ emissions.

There are normally no particulate removal systems on natural gas-fired power plants. According to figures from the US Energy Information Administration, the particulate emissions from natural gas combustion is roughly 400 times lower than for coal combustion, for the same unit of energy input, and so capture is not normally deemed necessary. Nevertheless, the combustion of natural gas, while it does not produce any fly ash, can still produce small amounts of particulates. Depending on their nature, some particulates from a natural gas-fired plant may be removed in a catalytic oxidation system (see below).

Oil-fired power plants will produce more particulate material than natural gas but less than coal. The oil combustion emissions are around 30 times lower than from a coal plant. However, heavy fuel oil plants can produce much larger levels. By way of an example, according to an emissions report compiled by the North American Commission for Environmental Cooperation, residual oil-fired power plants in Mexico were responsible for between 60% and 70% of all power plant PM_{10} and $PM_{2.5}$ emissions[4].

4.7 VOLATILE ORGANIC COMPOUNDS AND CARBON MONOXIDE

VOCs are organic substances that have a relatively high vapor pressure at ambient temperature and will therefore enter the atmosphere readily. Their specific definition varies from region to region; for example, the European Union defines a VOC as any organic compound with a boiling point of 250°C or lower.

VOCs are widespread. They are emitted from a variety of household products, and the concentration inside homes and buildings is often higher than outside. Not all VOCs are harmful but there are many that are considered so. In addition, VOCs can react with nitrogen dioxide in sunlight to produce ozone and a range of compounds that both contribute to photochemical smog and are a source of particulates. Small quantities of VOCs are generated during coal combustion, and they are also produced in natural gas-fired and oil-fired power stations. In both cases they are the result of incomplete combustion of the fuel.

[4]North American Power Plant Air Emissions, published by the Commission for Environmental Cooperation (CEC).

Carbon monoxide (CO) is also produced when fossil fuel combustion is only partially complete. As with VOCs, it can be found in the emissions from coal-fired power plants and natural gas-fired stations and well from power generation units burning oil. In a coal-fired power station the measurement of CO concentration can be used to determine the optimum combustion conditions and the right amount of air to be admitted to the boiler to achieve the most economical level of combustion. The optimum CO concentration in the flue gases is around 400 ppm[5] or around 460 mg/m^3.

Emissions of both VOCs and CO are regulated. Their removal from exhaust gases can normally be achieved using a catalytic oxidation system in which a metallic catalyst is used to promote the reaction of CO and VOCs with oxygen remaining within the flue gas stream, converting them into carbon dioxide. A catalytic oxidation system is normally positioned before any SCR in the power plant to prevent interaction between them.

4.8 HEAVY METALS

Coal can contain a range of metals, and these can be emitted in fly ash carried away in the flue gases of a power plant. A study in Poland, for example, found 120 ppm of zinc in fly ash from a power plant in Upper Silesia, together with 38 ppm of copper, 41 ppm of nickel, 44 ppm or lead, 64 ppm of chromium, and 3 ppm of cadmium[6]. The range and quantities of metals in fly ash will depend on the source of the coal, so it varies from region to region. Other potential pollutants includes arsenic and mercury, which can be released in vapor in hot flue gases.

In large power stations with efficient dust collection systems, most of the particles containing these metals will be captured. However, where capture efficiency is low these particles will be emitted into the environment.

[5]Carbon Monoxide Measurement in Coal-Fired Power Boilers, an application note published by Yokogawa.
[6]Heavy Metals in Fly Ash from a Coal-Fired Power Station in Poland, D Smoka-Danielowska, *Polish Journal of Environmental Studies*. Vol. 15, No. 6, pp 943–946, 2006.

The emission of mercury from coal-fired power plants has become specific concern in recent years. The metal is relatively volatile and so enters the gas-phase easily, particularly at elevated temperatures. Coal-fired power stations are considered to be the main anthropogenic source of mercury in the environment. Around 25% of that released from the coal will be captured by dust collecting systems although this can rise to as high as 60% if a wet sulfur dioxide scrubbing system is also installed in the plant. An SCR system can capture addition mercury from certain coals. For higher levels of capture, activated carbon particles can be injected into the flue gas stream. Metals such as mercury will become adsorbed onto their surface, and the particles are then collected in the dust capture system.

CHAPTER 5

The Nuclear Question

Nuclear power evolved from a variety of nuclear weapons programs during the 1930s and 1940s. It offered a new, advanced method of generating electricity, which was touted as the future for power generation with the prospect of limitless clean energy. After a slow start during the 1950s and 1960s, construction of nuclear power plants accelerated during the 1970s, particularly in the developed western nations. However, questions about safety led to a dramatic slowdown of nuclear construction from the 1980s onwards, a slowdown from which the industry has never recovered. Nuclear power production peaked during the first decade of the 21st century. At the end of 2015, there were 439 commercial reactors in operation.

Although growth has slowed down, the use of nuclear power raises important environmental questions and as with most environmental issues it is a matter of weighing advantages and disadvantages. The positive aspects of nuclear power include the fact that generating electricity from a nuclear station does not involve the release of any carbon dioxide into the atmosphere and so a nuclear plant can provide part of the solution to reducing global warming. Nuclear power can also provide energy security in countries that have limited natural energy resources. The negative aspects of nuclear power are all linked to the fact that nuclear generation is based on exploitation of nuclear reactions. These reactions produce a range of potentially hazardous waste products. In addition, the effects of a major accident at a nuclear power plant can be far-reaching. When there have been accidents, these have had a massive effect on the popular perception of nuclear power.

5.1 NUCLEAR POWER, NUCLEAR WEAPONS, AND NUCLEAR ACCIDENTS

Added to these considerations is another perceived link between nuclear and nuclear weapons. While the nuclear industry would claim

Electricity Generation and the Environment. DOI: http://dx.doi.org/10.1016/B978-0-08-101044-0.00005-6

that the civilian use of nuclear power is a separate issue to that of atomic weapons, the situation is not that clear cut. Nuclear reactors are the source of plutonium that can be used to make a nuclear weapon. Plutonium creation depends on the reactor design, and it is possible to build nuclear reactors that produce very little or no nuclear isotopes that are useful for weapons production. However the reactors that are in use today virtually all produce material that can be used for weapons. In addition, most nuclear power plants require enriched uranium and therefore rely on uranium enrichment plants. Highly enriched uranium is another material capable of being fashioned into a bomb. Both are therefore areas of international concern.

The danger is widely recognized. Part of the role of the International Atomic Energy Agency is to monitor nuclear reactors and track their inventories of nuclear material to ensure that none is being sidetracked into nuclear weapons construction.

The effects of the detonation of a nuclear weapon are devastating, as history has clearly demonstrated. Of course a nuclear power plant is not a nuclear bomb. Unfortunately for the nuclear power industry some of the after-effects of the detonation of a nuclear device can also be produced by a major civilian nuclear accident. The contents of a nuclear reactor core include significant quantities of extremely radioactive nuclei. If these were released during a nuclear accident, they would almost inevitably find their way into humans and animals via the atmosphere or through the food chain.

Large doses of radioactivity or exposure to large quantities of radioactive material kills relatively swiftly. Smaller quantities of radioactive material are lethal too, but over longer time scales. The most insidious effect is the genesis of a wide variety of cancers, many of which may not become apparent for 20 years or more. Other effects include genetic mutation that can lead to birth defects.

The prospect of an accident leading to a major release of radioactive nuclides has created a great deal of popular apprehension about nuclear power. The industry has gone to extreme lengths to tackle this apprehension by building ever more sophisticated safety features into their power plants. Unfortunately the accidents at Three Mile Island in the United States, Chernobyl in the Ukraine, and Fukushima Daiichi in Japan suggest that it may be impossible to build a nuclear power plant that is entirely safe. For modern plants, the risk of an accident may be extremely low. The difficulty is in persuading the public that any risk is acceptable when the stakes are so high.

Unfortunately the fear associated with nuclear accidents has recently been magnified by a rise in international terrorism. The threat now exists that a terrorist organization might seek to cause a nuclear power plant accident or, by exploiting contraband radioactive waste or fissile material, cause widespread nuclear contamination.

So far a peacetime nuclear incident of catastrophic proportions has been avoided, although both Chernobyl and Fukushima caused extensive disruption and in the case of the former a disputed number of deaths as a result of radioactive exposure. Smaller incidents have been more common and low-level releases of radioactive material have taken place. While these are rarely serious, they raise other issues.

One of these issues is the level of the danger from exposure to low radiation levels. The effects of low levels of radioactivity have proved difficult to quantify. Safe exposure levels are used by industry and regulators, but these have been widely disputed. On the one hand some would claim that there is no safe level of exposure. On the other hand there are natural sources of radiation to which everybody on the planet is exposed, so a level of exposure that is lower than that experienced naturally might be considered insignificant. Again, it is a matter of trying to establish risk levels and then to determine what level of risk is acceptable.

5.2 NUCLEAR POWER AND GLOBAL WARMING

One of the main advantages of nuclear power promoted by the nuclear industry today relates to its ability to provide low carbon electricity generation. According to the International Energy Agency (IEA), nuclear power was the largest source of low carbon electricity among the countries of the Organization for Economic Cooperation and Development (OECD) countries in 2013 with 18% of total electricity production. Across the globe as a whole, its share of production was 11%, making it the second largest contributor after hydropower[1].

Based on an IEA scenario for future power generation under which the rise in the global temperature is restricted to 2°C, the organization

[1]Technology Roadmap: Nuclear Energy, 2015 Edition, IEA and NEA, 2015.

suggests that nuclear generation would need to more than double from its present level, reaching 930 GW of installed capacity by 2050, when it would provide roughly 17% of global electricity production. This would represent an ambitious program of nuclear construction. However, it faces a number of hurdles.

Much of the existing nuclear capacity is in OECD countries. The members of the OECD are mostly rich, developed nations, and many of these invested in nuclear power in the early days of nuclear evolution. As a consequence, many of these countries have benefited from fleets of nuclear plants providing cheap base-load power. The cost of new nuclear power plants makes the technology inaccessible to all but the richest nations today yet growth in generating capacity is more urgent elsewhere. Small modular nuclear power plants might make the technology more accessible but such plants are not currently available commercially.

On top of that, the 2011 nuclear accident in Japan has curtailed much global nuclear activity while the construction of renewable generating capacity from wind and solar power continues to grow. There is already evidence that the low cost of these technologies is beginning to undercut others, particularly nuclear power, and this trend will continue. Meanwhile the challenging financial situation across the globe in the second decade of the 21st century makes it extremely difficult to find funding for capital-intensive projects such as nuclear power stations.

So, while nuclear power can contribute to reducing global warming, it will not be the first choice for new generating capacity in many, if not most parts of the world. Perceptions may change and if the cost of nuclear construction can be reduced by the availability of small, standardized nuclear power units then the appetite for nuclear power may improve. The danger for the industry is that nuclear technology will be overtaken by renewable developments elsewhere, and that when new nuclear technologies are available, nuclear growth will be difficult to justify economically.

5.3 RADIOACTIVE WASTE

As the uranium fuel within a nuclear reactor undergoes fission it generates a cocktail of radioactive atoms within the fuel rods inside the plant. Eventually the uranium fuel becomes exhausted. At this point the fuel rods will be removed from the reactor. They must then be disposed of in a safe manner. Yet after more than 60 years of nuclear fission, no safe method of disposal is widely available.

Several options are considered viable. The best large-scale method would appear to be the disposal of waste in underground bunkers built in stable rock structures. However while the principle has been agreed finding a site where construction can take place has proved extremely difficult. Reprocessing the waste fuel to remove and reuse the uranium and plutonium fissile material it contains is another option. This would reduce the volume of the residual waste, although the residue of high-level waste still requires secure disposal. Spent fuel reprocessing has been carried out in one or two countries but meanwhile in most countries there is no agreed solution to the growing backlog of nuclear waste. As a result most spent nuclear fuel has been stored in ponds at the nuclear power plants where it was produced. This is now causing its own problems as storage ponds designed to store a few years' waste become filled or overflowing.

Radioactive waste disposal has become one of the key environmental battlegrounds over which the future of nuclear power has been fought. Environmentalists argue that no system of waste disposal can be absolutely safe, either now or in the future. And since some radioactive nuclides will remain a danger for thousands of years, the future is an important consideration.

The quest to solve the problem continues. For many years underground burial has been the preferred option for the nuclear industry. This requires both a means to encapsulate the waste and a place to store the waste once encapsulated. Encapsulation techniques have already been developed. These include sealing the waste in a glass-like matrix that is then stored in heavy steel containers. The waste still generates heat, even in this form, and so must be cooled once it has been encapsulated. However the encapsulation should make it impossible for the waste to escape into the environment.

While encapsulation has been demonstrated, construction of an underground store has yet to be realized. An underground site must be in stable rock formation in a region not subject to seismic disturbance. Such locations have been identified and sites in the United States and

Europe have been studied for many years but none has been built yet. The most advanced project of this type is the Onkalo repository in Finland where excavation of the underground cavern has begun but the project awaits a construction license from the government. If this is granted, the first waste fuel is expected to be stored around 2020. However this site is designed for waste from Finnish nuclear plants. Other countries still have to find their own solutions.

Fuel reprocessing may offer another partial solution to the problem of nuclear waste. The reprocessing of spent fuel is part from the fuel waste and reused in nuclear fuel. However once reprocessing is complete there is still a significant residue that contains a variety of radioactive isotopes that are of no use in reactors. So, while reprocessing reduces the volume of waste it does not entirely solve the problem.

Other schemes have been proposed for nuclear waste disposal. It is possible to return the high-level waste containing radioactive isotopes to a reactor where they are bombarded with neutrons and where they eventually react to produce less harmful isotopes. This appears to be costly. Another involves loading the fuel into a rocket and shooting it into the sun. Yet another proposes utilizing particle accelerators to destroy the radioactive material generated during fission.

Unfortunately, while there are many proposals for the disposals of radioactive waste there are limited practical solutions available. In the meantime the volume of radioactive waste continues to increase and so does the environmental problem it represents.

5.4 WASTE CATEGORIES

Spent nuclear fuel and the waste from reprocessing plants represent the most dangerous of radioactive wastes but there are other types too. These come from a variety of sources. Anything within a nuclear power plant that has even the smallest exposure to any radioactive material must be considered contaminated. One of the greatest sources of such waste is the fabric of a nuclear power plant itself. This creates a large volume of waste when a nuclear power plant is decommissioned.

High-level wastes are expected to remain radioactive for thousands of years. It is these wastes that cause the greatest concern and for which some storage or disposal solution is most urgently required.

However these wastes form a very small part of the nuclear waste generated by the industry. According to the World Nuclear Association, the high-level waste only makes up around 3% of the total by volume. Most is the low-level waste. Even so it too must be disposed of safely. To deal with these different wastes, regulatory authorities have developed nuclear waste categories.

In the United States spent fuel and the residual waste from reprocessing plants is categorized as high-level waste[2] while reminder of the waste from nuclear power plant operations is classified as low-level waste. There is also a category called transuranic waste that is waste containing traces of elements with atomic numbers greater than that of uranium (92). All elements with a higher atomic number than uranium are naturally radioactive. Low-level wastes are further subdivided into classes depending on the amount of radioactivity per unit volume they contain.

In the United Kingdom there are three categories of waste: high-level, intermediate-level, and low-level. High-level includes spent fuel and reprocessing plant waste, intermediate-level is mainly the metal cases from fuel rods, and low-level waste constitutes the reminder. Normally both high- and intermediate-level waste require some form of screening to protect workers while low-level waste can be handled without a protective radioactive screen.

Low-level waste will often be disposed by shallow burial, often after compacting, and in some cases it may be incinerated in a special waste combustion plant to reduce the volume before burial. Intermediate-level waste needs have to be shielded because it contains higher levels of radioactivity. It may be sealed in concrete of bitumen before burial. However unlike high-level waste this type of waste does not need cooling when it is stored.

5.5 DECOMMISSIONING

A nuclear power plant will eventually reach the end of its life, and when it does it must be decommissioned. At this stage the final and perhaps the largest nuclear waste problem arises. After 30 or more

[2]The US Department of Energy does not classify spent fuel as waste, but the Nuclear Regulatory Commission does.

years[3] of generating power from nuclear fission, most of the components of the plant have become contaminated and must be treated as radioactive waste. This presents a problem that is enormous in scale and costly in both manpower and financial terms.

The cleanest solution is to completely dismantle the plant and dispose the radioactive debris safely. This is also the most expensive option. A half-way solution is to remove the most radioactive components and then seal up the plant for from 20 to 50 years, allowing the low-level waste to decay, before tackling the rest. Two Magnox reactor buildings in the United Kingdom were sealed in this way in 2011 and are expected to remain in that state for 65 years. A third solution is to seal the plant up with everything inside and leave it, entombed, for hundreds of years. This has been the fate of the Chernobyl plant.

Decommissioning is a costly process. Regulations in many countries now require that a nuclear generating company put by sufficient funds to pay for decommissioning of its plants. The US Nuclear Energy Institute suggests that the cost of decommissioning a US power plant is between $450million and $1.3 billion, based on figures from the US Nuclear Regulatory Commission from 2013. The US utility Southern California Edison has put aside $2.7 billion to decommission its San Onofre power plant, expecting this to cover around 90% of the total expenditure. Meanwhile in 2011 the UK government estimated nuclear decommissioning costs for its existing power plants to be £54 billion. When building a new nuclear plant the cost of decommissioning must therefore be taken into account.

---

[3]Nuclear plants in many parts of the world are now seeking operating license extensions to allow them to continue operations for up to 60 years.

CHAPTER 6

Renewable Energy and the Environment

Power generation from renewable energy resources is vital if the generation of electric power is to become sustainable and if damage to the environment caused by the combustion of fossil fuels is to be avoided. However renewable power generation technologies have their own effects on the environment. These, too, must be taken into account when planning new generating capacity.

The impact of the individual renewable technologies varies. Hydropower plants can cause major environmental upheaval, particularly where the scheme includes a reservoir as can tidal power. Biomass power plants require a biomass fuel crop which may take up land that would otherwise be used for food crops. There are also combustion emissions associated with most biomass power plants. Solar and wind power are both relatively low-intensity generation sources that must be deployed over large areas if they are to provide sufficient energy to replace more traditional source of electricity. This can be disruptive locally. Marine sources of renewable energy will also occupy large areas at sea if they are deployed on a similar scale to wind and solar power. Geothermal power generation is relatively benign but exploitation can produce local pollution if the hot brine extracted from underground is not returned to the reservoir.

There is a further implication when the large-scale intermittent renewable resources such as wind and solar power are used extensively. These sources require some form of backup so that when there is insufficient renewable electricity being generated, grid control centers can replace the shortfall with an alternative to maintain the supply. This backup is normally based on some type of fossil fuel power plant, often a natural gas turbine power station. Such a plant will generate emissions when it is in use. Furthermore this type of power plant operates with less efficiency and generates more emissions when it is required to change its output regularly to support renewable

Electricity Generation and the Environment. DOI: http://dx.doi.org/10.1016/B978-0-08-101044-0.00006-8

generation than it would if it was operating under steady-state conditions, as it might if it was the main source of electric power.

6.1 THE TECHNOLOGIES

Table 6.1 shows the global installed generating capacity for each of the main renewable technologies in 2013, based on figures collated by the International Energy Agency (IEA). Hydropower was the largest in terms of installed capacity with 1136 GW. Hydropower has been a significant source of electricity since the beginning of the 20th century and continues to play a major role in global generation.

The combustion of biomass is also an old established means of power generation. The Bioenergy entry in the table primarily covers direct biomass combustion but also includes some other sources such as the generation of combustible gas from waste. Together these contributed a further 88 GW in 2013. Most biomass capacity uses combustion technology that is similar to that employed by coal-fired power plants.

The other technologies in the table are generally considered to be new renewable technologies. Interest in these began to develop strongly after the oil crises of the 1970s but it was not until the beginning of the 21st century that any of them became firmly established. Since then two have grown rapidly. Wind power, with 303 GW of installed

Table 6.1 World Renewable Generating Capacity by Type, 2013[1]	
Technology	2013 (GW)
Hydropower	1136
Bioenergy	88
Wind	303
Solar photovoltaic	136
Solar thermal energy	4
Geothermal	12
Marine energy	1
Total	1680
Source: *International Energy Agency*.	

[1]Medium-Term Renewable Energy Market Report 2015, International Energy Agency, 2015.

capacity in 2013, has seen the most rapid growth. Solar photovoltaic generation (solar cells) has also advanced strongly and there was 136 GW of capacity in 2013. Solar thermal power technology has developed more slowly but there was 4 GW of solar thermal capacity at the end of 2013 while marine energy contributed a further 1 GW. Geothermal energy, a resource that is limited geographically in its availability, provided 12 GW in 2013.

Hydropower, the most successful renewable resource in terms of installed capacity, exploits flowing water to drive turbines and generate power. Large hydropower plants can be up to several thousand megawatts in generating capacity, making them among the largest power plants in existence. Construction of such a plant is an enormous engineering project, often involving the erection of a massive dam and the inundation of a large area behind the dam to form a reservoir. This is usually environmentally disruptive but good management of the construction phase can ease the problems. Smaller hydropower schemes are less intrusive.

Wind power is the fastest growing of the new renewable technologies. Most wind turbines are installed on land but there is a small but growing offshore capacity. Wind farms, comprising arrays of wind turbines, can cover a large area but they are normally designed so that the land beneath them can still be used for agriculture. Wind power is an intermittent source and requires supplementary energy sources to provide continuous power. These may include some form of energy storage system.

Solar power comes in two forms, solar photovoltaic (solar cells) and solar thermal. The former is the second most successful new renewable energy source after wind power. A large solar power plant needs to collect solar energy over a large area, so utility scale solar power plants are physically intrusive. However solar cells can also be deployed on rooftops where they have little environmental impact and are rarely intrusive. Like wind power, solar energy is intermittent.

Biomass energy for power generation comes in a variety of forms. Small-scale plants often utilize some form of waste material but large-scale biomass power generation requires large volumes of biomass fuel and this must be grown on special plantations. In some cases this can lead to displacement of food crops. In addition, the combustion of biomass does lead to the production of some pollutant emissions that may need to be controlled in the same way as the emissions from coal or natural gas-fired power plants.

Marine energy sources include tidal power and tidal current power plants. A traditional tidal power plant with a barrage that traps tidal water and uses it to drive turbines—operation is similar to that of a traditional hydropower plant—can cause a change to the environment in which it operates. Alternative designs such as offshore tidal lagoons are less intrusive in this respect. Systems that are designed to exploit the tidal flow directly and those that exploit ocean currents are like underwater wind turbines and their impact will depend on the number of units deployed at any particular site.

Geothermal energy exploits hot brine from underground reservoirs as a source of heat that can be used to drive a heat engine. The latter may be a steam turbine or some other type of closed cycle turbine system. Provided the brine that is extracted from the reservoir is returned into the reservoir after use, then a geothermal power station should have a low impact on its surroundings. Depending on the size of the plant compared to the size of the reservoir being exploited, the heat content of the underground resource may become depleted over time. Some geothermal plants release carbon dioxide from the reservoir into the atmosphere.

6.2 MANAGING SOLAR AND WIND ENERGY INTERMITTENCY

Two of the most important renewable energy sources, solar power and wind power, are intermittent. This means that neither of them can provide a continuous source of electricity. To guarantee security of supply, both require a supplementary source of electricity that can step in when the intermittent source is not available. However the provision of backup for renewable energy can be costly depending upon the generating source used.

Coping with intermittency while maintaining security of supply is one function of the grid system control center. The control center will have various tools at its disposal. These include demand management facilities that allow the supply to some consumers to be reduced when the supply falls. Another important tool is accurate forecasting to predict wind and solar output for the day or week ahead. Alternative supplies can then be scheduled in advance for periods of low renewable output. This will reduce costs by providing advanced notice to the backup generators and allow slower acting plants to respond. Fossil fuel power plants that can ramp up their output rapidly and at short notice are often the most expensive to operate.

One of the most cost-effective methods of supporting intermittent renewable generation is by using hydropower. Hydropower plants equipped with reservoirs can offer a quick responding source of backup power. Reservoirs are a form of energy storage and turbines can be switched on and off at will in such plants without wasting energy. Response time can be a matter of seconds too, so they can step in quickly if necessary. Another method of supporting intermittent renewable energy while avoiding fossil fuels is with one of a range of dedicated energy storage plants. However energy storage technologies are still relatively expensive and grid connected storage is limited in most parts of the world.

When hydropower and energy storage facilities are not available, the main alternative is to use a fast acting fossil fuel-fired power plant. This will most often be a gas turbine-based station burning natural gas, but modern coal-fired plants may be able to perform the same function. Some modern nuclear plants may be able to provide this type of service too. The requirement is that the plant should be able to be brought into service and taken out of service quickly and that it is capable of ramping its output up and down rapidly.

From an environmental perspective, it would be preferable to be able to avoid the need for fossil fuel-fired backup plants. This is feasible but will require some major changes to grid configurations and the way the grid is operated. An important aspect of this is likely to be the addition of much more energy storage capacity. Already today, energy from wind and solar power plants is being shed (i.e., wasted) by plant operators and grid managers when there is no demand for it. With sufficient storage capacity, this energy can be stored and then made available when renewable output falls because there is no wind or night has fallen. Energy efficiency can also play its role by reducing the overall demand.

Careful renewable planning is also vital. For example, wind turbines can only generate electricity when the wind blows, so a lone turbine or single wind farm will be at the mercy of local wind conditions. However when the wind is not blowing in one part of a region or country it is often still blowing in another. If wind energy is averaged over the whole region instead of being assessed, wind farm by wind farm, then it becomes a much more reliable resource. Different renewable sources can complement one another too. Considering solar and wind energy, solar energy is most active during the day and in the summer; wind power is generally greater during the winter and the wind in some regions is stronger at night. Of course, biomass energy, if it is available, can step in at any time. When all these are combined with hydropower in a portfolio of renewable capacity, the result is a much more reliable source of energy. For today, however, additional fossil fuel backup remains vital.

As earlier chapters have shown, the combustion of fossil fuels leads to the release of a range of pollutants into the environment. These are minimized by using emission control systems. However the problem is exacerbated when this type of power plant is operating in support of intermittent renewable power sources because the combustion plant will not be operating under highly controlled steady-state conditions. Instead its output will be varying to compensate for varying shortage of renewable energy. When combustion plants operate under non–steady-state conditions, they are less efficient—so the energy they produce is more costly—and they generate more pollutants. Controlling emissions becomes more difficult under these conditions too.

6.3 ROTATING MACHINES AND GRID INERTIA

In a traditional grid system there are large a number of fossil fuel-fired, hydropower, and nuclear power stations providing power to a national grid system that stretches across a whole region or country. All these traditional generating units that are connected to the system are massive rotating machines. The aggregate rotating mass of these rotating machines provides the grid with an enormous rotational inertia. If there is a sudden major fault on the system, the supply does not fail immediately because this inertia means that the machines will continue to turn as they slow, giving grid controllers time to step in and maintain stability.

Most renewable generation systems do not provide a similar rotational inertia. With the exception of large hydropower generators, the main renewable resources either have no rotational energy (solar cells) or are connected through an electronic power converter that makes any rotational inertia invisible to the grid.

As the amount of renewable energy increases, this loss of rotational inertia will tend to make grid systems more sensitive to disruption. One solution is to add physical rotating machines to the grid, usually large generators that do not generate energy but simply spin at the grid frequency. Another is to break the supply system down into a patchwork of interconnected mini grids, each of which can operate independently on the whole if necessary while being able to provide support the whole grid if called to do so.

Meanwhile, new renewable generating sources are being asked to supply some form of grid support to counter this loss of rotational inertia. However the loss of inertia remains on the important unsolved issues for a future sustainable electricity supply.

6.4 BIOMASS ISSUES

There are a range of methods of generating electricity from plant material but most are only relatively small in scale. It has been traditional to burn waste materials, particularly wood waste, in combustion plants that generate steam for steam turbines. Wood waste can come from forestry management or from industries that process wood. There are also a wide range of agricultural wastes that can be burned in the same way, for example the residue from the harvesting and processing of sugar cane is commonly used to raise energy. Straw from corn crops can similarly be burned, as can the husks left from rice processing and coconut husks.

Another way of using biomass as an energy source is to ferment it in the absence of air when it will produce a methane-rich gas that can be burnt. This process takes place in landfill waste sites across the world and such sites generate large volumes of methane that is normally captured today and used to generate electric power. Animal wastes can be fermented too, and this is another small-scale source of energy.

None of these is capable of providing sufficient fuel for large-scale generation of electricity. That requires crops that are grown specially as combustion fuel. There are several such crops available today. Fast growing trees are probably the most popular but grasses are also potential energy crops and these are capable of growing rapidly too, particularly on ancient grasslands such as the Prairies or the Steppes. There are also crops that can be used to produce oil. Such oils are more generally used as alternative transportation fuels but can be used for power generation. Ethanol production is popular in some parts of the world but again the fuel is mostly used in vehicles.

While there are a range of habitats that might be used to raise energy crops in a sustainable manner, today it is more likely either that energy crops will be grown on existing cropland or that natural biomass resources will be exploited. Where energy crops use agricultural land, there is a danger that the energy crop will compete with alternative food crops, potentially reducing the availability of food and making the food that is more expensive. The problem is likely to be worst in countries of the developing world where food is already in short supply while an energy crop offers a potentially lucrative business opportunity.

Meanwhile there is evidence that a growing popularity for biomass in Europe is leading to the exploitation of forests in places such as Canada, from which wood chips are produced for export. Development of another energy crop, oil palm, has been blamed for rain forest clearance in some parts of Asia.

In addition to land usage issues, there is also a question about the sustainability of biomass as an energy source. Biomass fuels are exclusively combustion fuels and when they are burnt they generate carbon dioxide that enters the atmosphere in exactly the same way as a fossil fuel would generate carbon dioxide. The rationale for the sustainability of biomass is that when a biomass crop is grown, it will do so by absorbing carbon dioxide from the atmosphere. When it is burned, this carbon dioxide returns to the atmosphere but then the crop is regrown, taking up the carbon dioxide again. In this way biomass is environmentally neutral, neither adding nor subtracting to the atmospheric burden of carbon dioxide.

For this to be effective, there needs to be strict biomass accounting. The regular growth and cropping of grasses and fast growing trees are easy, in principle, to keep account of but if the biomass comes from natural sources such as forests then it becomes much more difficult. In addition, when biomass is being traded from continent to continent there are questions about the transportation energy costs too.

This accounting can be complicated by one of the mechanisms that has been used in support of global greenhouse gas control that was built into the Kyoto Protocol. The mechanism allows a company or country to offset its carbon dioxide emissions in one place by sponsoring its removal elsewhere, such as by the reforestation of a remote area. While this type of scheme is attractive, it is full of loopholes when it comes to ensuring that the amount of carbon dioxide being generated is truly being offset and over what timescale.

The final problem with biomass is simply that it is a combustion fuel. Whatever its type, when it is burnt to generate energy it will also produce airborne pollutants and other residues. Of course its chemical composition is different to that of the equivalent fossil fuel and this leads to somewhat different emission profiles. For example, biomass crops contain little sulfur and so they do not generate significant quantities of sulfur dioxide when burnt. However they can produce particulates material, organic residues, and carbon monoxide. In addition there is likely to be production of some nitrogen oxides. The release of emissions from a biomass power plant may be exacerbated by the fact that many such plants are small and have low efficiencies. In addition, it may not be cost effective to introduce the same emission control technologies that might be used in a larger fossil fuel-fired plant.

6.5 HYDROPOWER

Hydropower is one of the oldest sources of electrical energy and it continues to be important for global power generation, producing around 15.4% of all the electricity generated in 2014, as noted in Chapter 1. This makes it the largest single source of renewable energy in operation.

Hydropower plants are usually categorized by size. The main division is into large and small hydropower, with the small category including mini and micro schemes. The division between small and large varies from country to country but is typically between 10 and 30 MW, as shown in Table 6.2. Small hydropower then encompasses plants between the upper limit and from 1 MW to 100 kW schemes are classified as mini hydropower and below 100 kW is micro hydropower.

In recent years large hydropower plants have been considered apart from the other categories and they often do not attract the same subsidies or incentives as other renewable power plants. Small, mini, and micro hydropower schemes, on the other hand, can be eligible for such incentives.

For large hydropower there are two main types of plant, a dam and reservoir scheme and a run-of-river design. The dam and reservoir

Table 6.2 Hydropower Plant Categories[2]	
Micro	1–100 kW
Mini	100–1 MW
Small	1–10–30 MW
Large	>10–30 MW
Source: *UNDP/World Bank*	

type of development, as its name implies, involves construction of a dam across a river behind which a body of water, the reservoir, collects. This reservoir acts as a form of energy storage. Water will build up during the wettest season of the year and then it will slowly be depleted during the drier seasons. A run-of-river project has a small dam or barrage but not reservoir. Instead, it relies on the flow of water passing down the river as its source of energy.

By their nature, dam and reservoir schemes are much more intrusive and environmentally disruptive than run-of-river schemes because they inundate a large area of land. Everything within the area of the reservoir will be underwater for at least part of the year. This make the area uninhabitable; humans and animals will be displaced and all the existing plant life will be destroyed.

Large hydropower schemes can be extremely large. The largest in the world, the Three Gorges hydropower plant on the Yangtze River in China has a generating capacity of 22,500 MW. Its reservoir is 600 km long. The construction of this power station required the displacement of more than 1 million people and flooded 13 cities, 140 towns, and 1350 villages. A hydropower project of this scale will have a major effect on the region. In addition to the displacement that takes place, it may reduce biodiversity if it destroys the habitat of a rare species, plant, or animal.

Not all large hydropower schemes are quite so disruptive as the Three Gorges but they all present similar challenges. In addition to the displacement of people and animals, cultural sites of importance can be drowned. The weight of water in the reservoir can generate seismic tremors and a large reservoir can, ironically, produce large amounts of greenhouse gas.

[2]Private mini hydropower development study: the case of Ecuador, UNDP/World Bank, 1992.

The greenhouse gas is methane and it is generated if the trees and plants that grew in the region of the reservoir are not removed. If they are left, the conditions they experience underwater, particularly if there is little oxygen in the water, leads to anaerobic fermentation and decomposition—exactly the same process that takes place in a landfill waste site—and methane is one of the products. For most projects, this methane production will peak during the first 3–5 years after inundation and after 10 years the emissions should be similar to those from a natural lake. However there have been schemes built in which the reservoir is so wide and shallow that the amount of plant material underwater is enormous and methane production is massive, potentially making the hydropower plant more damaging to the environment than a similarly sized fossil fuel plant.

There are other issues too. A hydropower dam will stop sediment flowing down a river. This leads to two problems. If the river carries a large quantity of sediment, it may all settle in the reservoir that will eventually become full of sediment. It will then no longer be a reservoir. The second problem is that some downstream habitats rely on receiving the sediment to thrive. If a habitat is deprived of the sedimentary flow, then it will be changed or destroyed. The most high profile example of this is the Aswan High dam that was built on the River Nile in Egypt. The Nile delta has relied for centuries on sediment carried down by the river for its richness and for centuries it has provided a valuable agricultural resource. After the dam was built and sediment stopped flowing, erosion of the delta increased rapidly and fertility of the soils decreased necessitating artificial fertilizers to be used extensively.

In addition, a dam affects the flow of water downstream. Again, this may be vital to downstream communities that will suffer if deprived of it. This can be a particularly difficult problem when a river flows through several countries. When the River Euphrates was dammed in Turkey, communities in both Syria and Iraq experienced much lower flows. This was the cause of several interregional disputes.

Small, mini, and micro hydropower plants are generally much less disruptive than their larger relatives are. While a few small hydropower plants in the 5–30 MW range are based on a dam and reservoir design, most are run-of-river and only require a small barrage to create a head of water that can be exploited for power generation. Mini and micro hydropower developments often use novel turbines and designs. These are normally environmentally benign and have little local impact.

6.6 WIND, SOLAR, AND MARINE TECHNOLOGIES

Wind, solar, and marine technologies share in common that they are relatively low-intensity energy conversion systems when compared to fossil fuel and large hydropower projects. The largest wind turbines are approaching 10 MW in generating capacity but most are smaller. For large-scale generation, tens or hundreds of individual units are required to match the capacity of a central fossil fuel power plant and these will be spread over a large area because otherwise each wind turbine would interfere with its neighbors.

Solar power is similarly low intensity. While the amount of solar energy that reaches the surface of the earth is enormous, collecting enough for a high-capacity power plant requires a large area. However in this case the solar collectors will cover most of the available area, with little scope for agriculture.

Marine technologies mostly fit into a similar category. The exception is tidal power. When this involves building a barrage across a river to dam it and then using the barrage to collect water on the incoming tide, releasing as the tide goes out through turbines to generate power, the impact is likely to be similar to that of a large hydropower plant and dam. The ecology of the regions both behind and in front of the barrage will be changed significantly. The other marine technologies might be considered to be underwater "wind" turbines. They are generally smaller than wind turbines for the same generating capacity because moving water contains more energy per unit volume than moving air, but ecological considerations are similar.

The main problem with wind turbines is their visual impact. Not everybody likes to see them marching across the countryside. This can create difficulties with permitting. Offshore wind farms are less visually intrusive, particularly if they are out of sight of land, but even they can be subject to objections. Wind turbines generate noise too, but this is normally slight except at very close distance. Turbine blades may injure or kill animals, particularly birds and bats. While the danger in most cases is small, there are instances where large populations of a particular species near a wind farm can lead to increased risk.

Large or utility scale solar photovoltaic power plants can be visually intrusive but they are generally built at ground level and so are not visible from a distance. However they will occupy a large area of land. Typically a solar photovoltaic power plant will occupy between 1 and 4 ha/MW. This land cannot be used for raising crops. In many cases, solar plants can be located in scrubland or desert that has no value as cropland but landowners may decide to replace crops with solar energy harvesting. This has been seen, for example, in some parts of California[3]. Smaller solar photovoltaic installations, often on the rooftops of homes or commercial buildings, are generally less intrusive. Solar thermal power plants also require large areas of land, typically 2–5 ha/MW. Again, these will normally be constructed in areas of high solar insolation that are not used for raising crops.

Marine power plants (excluding tidal barrage style plants) have no visibility problem because they are sited underwater. The technology is still new and relatively untried and there are no large-scale plants of this type in operation. The impact of large-scale deployment has yet to be evaluated.

6.7 GEOTHERMAL POWER

Geothermal power is based on the exploitation of underground reservoirs of hot brine. This brine is pumped to the surface from underground and the heat energy in the brine is then used to drive a heat engine. The design of the energy extraction system will vary but the principles are broadly the same.

The environmental impact of this type of plant depends on its management. If the extracted brine, once exploited for energy production, is pumped back into the reservoir from which it was taken, then the overall impact should be minimal. If the brine is not returned or if only a part of it is returned, the remainder will likely enter local streams, rivers, or groundwater sources where it will pollute the water and make it unusable for drinking or irrigation. Depending on the scale of the problem, this may affect a small or a large area. In addition, some geothermal brine contains carbon dioxide and this is often released during operation of the plant.

[3]Solar Energy's Land-Use Impact, Carnegie Science, October 19, 2015.

Because the geothermal power plant takes energy from an underground reservoir, it must reduce the amount of energy within the reservoir. In practice this means it reduces the heat content and the temperature within the reservoir gradually falls. This can eventually lead to the reservoir no longer being suitable to support power generation. Because the reservoir receives its heat from within the earth, the temperature will rise again slowly. It is therefore possible to balance the size of a geothermal plant to that of the reservoir such that the heat extracted for power generation is replaced on roughly the same timescale and there is no overall depletion.

CHAPTER 7

The Hydrogen Economy

If the use of fossil fuels is to be abandoned over the course of this century, an environmentally desirable target but one that is not certain, then a range of technologies and processes that depend on fossil fuels will have to be abandoned. These will include all the common combustion technologies used in power generation as well as the use of reciprocating engines for transportation and natural gas for heating and cooking.

There is an alternative and that is to replace the fossil fuels in use today with different fuels which does not harm the environment when they are burnt. These could be one of a range of fuels such as ethanol and oils that are already being produced from crops, but none of these is likely to be able to replace fossil fuels in all applications. However there is one fuel that conceivably might do that, hydrogen.

Hydrogen, in gaseous form, will burn like natural gas. As such it can act as a replacement fuel for most of the applications that rely on natural gas today including gas turbines and reciprocating engines for power generation. Systems need adapting because hydrogen burns in air at a higher temperature than natural gas. This is normally solved relatively simply by diluting it with air or exhaust gases from the plant. More importantly, the combustion of hydrogen generates only water vapor. There are no carbon emissions.

Hydrogen can also be used as a transportation fuel. Hydrogen vehicles are already available and hydrogen distribution networks for vehicle refueling are increasing. Moreover, hydrogen can be transported in pipelines or it can be liquefied, creating a high-energy density form that can be moved in bulk. Hydrogen can also be used as a form of energy storage, using excess electrical power to produce it and then using the hydrogen to provide more electricity as required.

Where would this hydrogen come from? The simplest and cleanest solution is to produce it from water using electricity via a process

Electricity Generation and the Environment. DOI: http://dx.doi.org/10.1016/B978-0-08-101044-0.00007-X

called electrolysis. Provided the electricity is generated from a clean source, the hydrogen is also environmentally benign. Depending on demand, this would require vast quantities of both electricity and water and the process is relatively inefficient. However, if solar and wind power become cheap enough, then this may not prove a problem.

Whether this is practical on a sufficiently large scale to replace fossil fuel is not yet clear. The energy cost for producing, handling, and transporting the hydrogen fuel would be large. In many cases it would be much more efficient to use the electricity directly than to convert it into hydrogen that is then turned back into electricity somewhere else. In addition, there are alternative synthetic fossil fuels that could be made using carbon dioxide captured from power plants burning synthetic fossil fuels. Synthetic fossil fuel may well be more economical to produce and use than hydrogen.

7.1 HYDROGEN PRODUCTION[1]

If hydrogen is to be used as a substitute for natural gas, then an abundant source is required. Fortunately one is available, water. Water can be converted into hydrogen and oxygen by applying an electrical voltage to it using electrodes placed in water. When this is done and a sufficiently high voltage applied, hydrogen is generated at one electrode and oxygen at the other. The gases can easily be collected, separately, with the hydrogen stored and the oxygen released into the atmosphere again.

The electrolysis of water is a relatively efficient process with modern electrolyzers capable of around 90% efficiency and systems under development aiming for 94%. Converting the hydrogen back into electricity will be most efficiently carried out with a high-efficiency fuel cell. The best of these might be capable of 60% efficiency today, but future hybrid fuel cell configurations may achieve 75% efficiency. This would suggest a best round trip efficiency of 71% or an energy loss of 31 percentage points.

There are other potential sources of hydrogen. Today most of the hydrogen used by industry is made by reforming natural gas. This

---

[1]More details about hydrogen production and storage can be found in Hydrogen Production and Storage, R&D Priorities and Gaps, International Energy Agency, 2006.

releases carbon dioxide into the atmosphere but the carbon dioxide could feasibly be captured and sequestered to prevent it from adding to the carbon in the atmosphere. Coal can also be converted into hydrogen. This is less efficient in terms of energy, but coal is cheap. However it is unlikely that either of these represents a viable clean energy alternative.

Potentially more attractive environmentally is the use of methanol or ethanol. Both of these can be produced from biomass. Sugarcane and maize are both used extensively to manufacture ethanol today. Once produced, the ethanol or methanol can be reformed to make hydrogen. This will involve the production of carbon dioxide but if the cycle of growth, ethanol manufacture, reforming, and combustion followed by growth of a new crop is continuous, then the overall process should be carbon neutral. Another biomass source is wood that can be gasified and reformed into hydrogen in a similar way to coal.

The electrolysis of water to produce hydrogen will be clean if the electricity comes from renewable sources such as hydropower, solar power, or wind power. There may, in addition, be methods of generating hydrogen directly from sunlight without the need for electricity. Research into photoelectrolysis using solid-state electrodes has been underway for around 40 years. The principles of this process are well understood, but it has proved difficult to develop electrodes that can provide a commercially attractive source of hydrogen. A second direct approach is photobiological electrolysis (or biophotolysis). Biophotolysis is a two-step process. First, sunlight is absorbed by a plant and used to produce algae. The algae then use further sunlight and organic catalysts called hydrogenases to generate hydrogen. This process is at an early stage of development.

Finally, water can be split into hydrogen and oxygen at very high temperatures. The process occurs at around 3000°C. This can be achieved using sunlight if it is concentrated, allowing the direct thermal production of hydrogen. Again this is a process in an early stage of development.

7.2 HYDROGEN STORAGE AND TRANSPORTATION

It is important to have a simple and convenient means of storing hydrogen. Hydrogen can be stored in the gaseous, liquid, or solid phase. The simplest method is to store the gas at ambient temperature, under pressure in a lightweight steel or composite tank. Storage pressure can be as high as 700 bar. Compressing the gas for storage has an energy cost and the storage cylinders capable of withstanding this pressure are expensive.

Another method is to cool the gas cryogenically to just above its point of liquefaction. This increases the energy density significantly, but again there is an energy cost for cooling the gas. The storage vessel must be able to maintain the low temperature and this makes it costly too.

There is another, novel method of storing the gas using tiny glass microspheres. The microspheres are filled with the gas by heating them to around 300°C when they become permeable so that the gas can penetrate the spheres. They are then cooled to ambient temperature, and the gas becomes locked within them. The charged spheres can they be used, for example, as vehicle fuel by loading them into a fuel tank. The gas is released again by heating the spheres once more. This method is at an early research stage.

If hydrogen is cooled to $-253°C$ or around 20°C above absolute zero, the gas liquefies. Hydrogen can be stored in this form provided a cryogenically insulated vessel is available. This form of storage was used in the US space program to provide fuel for fuel cells in spacecraft and the space shuttle. Another method of liquid storage is in the form of sodium borohydride, $NaBH_4$. The compound will release hydrogen when it is reacted with water. While it is potentially a high-energy density method of storage, it is expensive because the cost of manufacturing the borohydride is expensive.

The third alternative is to store the gas in some solid-state form. This involves absorbing or adsorbing the gas into (or onto) a solid from which it can easily be released again. One option is to use a high surface area material that will adsorb and release the gas easily. Several materials are being examined but no strong candidate has yet emerged. The other option is to use metal hydrides, solids that will reversibly absorb or dissolve the gas. A number of promising materials of this sort have been identified and they offer one of the best options for future hydrogen storage for transportation applications such as hydrogen cars.

Bulk transportation of hydrogen will depend upon the application. It is possible to use pipelines for hydrogen transportation in the same way as natural gas but costs may be higher. Alternatively if high-density storage can be developed, it may be cost-effective to transport it using some form of bulk carrier.

7.3 HYDROGEN AND POWER GENERATION

Hydrogen can be used to generate electricity in many ways but the most efficient is likely to be using fuel cells. These devices are a little like batteries but with fuel supplied externally rather than from within the cell. A fuel cell requires hydrogen at one electrode and air or oxygen at the second, and it will generate electricity cleanly and with high-efficiency. The best simple fuel cells can probably achieve 60% efficiency. More complex hybrid cells may be able to achieve higher efficiency.

Aside from the fuel cell, hydrogen can be burnt in the same way as natural gas to generate heat for electricity production. The simplest combustion plant would be either a reciprocating engine or a gas turbine but combustion boilers that are designed to raise steam for a steam turbine could use hydrogen too. Most types of combustion plant would need some modification to adapt them to changed combustion conditions.

Whether the widespread use of hydrogen for power generation would make economic sense in a future energy economy will depend upon how that economy evolves over the next century. It may be more effective to use electricity from renewable sources directly where possible, only converting it into stored energy in the form of hydrogen to provide fuel for situations where a grid electricity supply is not appropriate.

CHAPTER 8

Low-Level Environmental Intrusion

The preceding chapters have been concerned with specific environmental effects associated with particular types of power generation: carbon dioxide emissions from fossil fuels, visual intrusion from wind farms, or population and wildlife displacement caused by hydropower development. However there are a range of environmental effects that all types of power-generating stations share. These are low-level effects such as noise, heat, or disruption caused by traffic movements associated with the generator. These are generally local in nature but this does not render them less intrusive for the populations they affect. These will be considered briefly in this chapter.

8.1 POWER PLANT CONSTRUCTION

The construction of a power plant will always have a local impact, however large or small the plant being built. The smallest generators, perhaps a few solar panels installed on the roof of a domestic dwelling, will probably cause little more than some unusual noise and one or two additional vehicle movements over 1 or 2 days. After the installers have gone, there will be little cause for concern unless the solar panels themselves are visually intrusive. Small rooftop wind turbines, if they are permitted in an urban area, will have a similar impact.

When it comes to the construction of a larger plant, then there will be much more local impact. Large power stations are not normally built in urban areas but there will often be some population in the area in which the plant is constructed. For this population there will be an increase in road traffic as components and building materials are delivered to the site and workers travel to and from the growing plant. The heavy vehicle movement will generate additional local pollution and noise and there will be more noise from the construction site itself. There may also be emissions from machinery at the site and depending on the size and type of plant being built there might be effluent flows into local streams or entering the groundwater. A large coal-fired

Electricity Generation and the Environment. DOI: http://dx.doi.org/10.1016/B978-0-08-101044-0.00008-1

power plant can take 3—4 years to build and the intrusion will continue throughout this period. Nuclear plants can take even longer, some exceeding a decade of construction.

Wind turbines and wind farms present their own problems. Modern wind turbines for utility-sized wind farms are generally massive devices with towers up to 100 m tall and turbine blades often approaching a similar length. The best wind sites are usually in remote rural locations with limited transportation infrastructure. Transporting the components of these large wind turbines, and there may be tens or even hundreds in a single wind farm, can be a major logistical problem that will have an impact on all activities in the area.

Depending on the site and size of a new power generator, another potential source of disruption is the connection of the plant to the grid. Fossil fuel and nuclear power plants will normally be built close to transmission lines and connection is relatively straightforward with a short interconnection between plant and grid. However for large renewable plants built in locations remote from the backbone of a grid system, a new, long transmission line may be required. This may require right of way across a rural landscape and if it is over-ground it will add a new local feature.

8.2 POWER PLANT OPERATION

Once a power plant has been built the disruption associated with its construction will cease but there will still be a range of activities relating to operation of the plant. All power plants require regular maintenance if they are to remain operational and this will occasion regular, if infrequent traffic. Fossil fuel plants have to be supplied with fuel. For a gas plant this may be via a pipeline but coal and oil plants will often have fuel transported using bulk carriers, sometimes by rail and at other times by road. The same is true of biomass combustion plants.

An operating power plant may produce some noise, although this will generally be very local. Airborne emissions might cause smells. There could also be spillages ranging from fuel to lubricating oil and coolant. If these are not contained, they may affect the surrounding area, for example by entering groundwater or local streams or rivers. There may also be occasional failures. Wind turbine blades sometimes shear and fall and towers might come down. Combustion boilers might explode. Sometimes a hydropower dam will fail. Such occurrences are rare but they do happen.

Combustion plants that use heat to raise steam for a steam turbine require some form of cooling. This is necessary to create as large a temperature difference between the inlet and the outlet of the steam turbine as possible. The larger the temperature difference, the more efficient the heat engine. Some plants use air-cooling. This will be common where ambient air temperatures are generally low or where there is little or no water available. When air is used for cooling, it becomes hotter as it carries away heat from the process. This will have an impact on the local environment. It may attract different species of bird or insect, for example, or enable unusual plants to grow. Large thermal power plants, including both coal-fired power stations and nuclear power stations, more usually use water for cooling. This will be taken from a local river or in some instances from the sea. As with air-cooling, water cooling results in water being returned at a higher temperature than normal. Again this can encourage different species that take advantage of warmer water temperatures.

8.3 POWER PLANT DECOMMISSIONING

When the useful life of a power plant is over, the plant must be decommissioned. This is the reverse of construction but like construction it will involve additional traffic movement, noise, and some extra local pollution. The most difficult power plants to decommission are nuclear power plants because components and much of the fabric of the plants will have been exposed to radiation and must be handled carefully. Conventional plants present fewer problems but they can still be massive undertakings.

Power plant sites are often valuable because they are approved sites for power generation. Owners of such sites may use the opportunity to build a new plant. The lifetime of a large fossil fuel–fired plant or nuclear plant will be 30 years or more and with adaptation, the plant life may easily be extended for longer. Under these circumstances there may be no obvious decommissioning. Where technology is advancing rapidly, such changes may be more rapid. Wind technology can change dramatically over 10 years and it may be economical to take down turbines that are 10 years old, or sometimes younger than this, and replace them with larger, more efficient modern turbines. This will incur the same environment disruption that the original construction involved.

These types of environmental disruption may appear minor when viewed from a regional or national perspective. However they can be extremely intrusive locally. They may only affect a tiny part of the population but that tiny part of the population may have its way of life turned upside down. Balancing this against a perceived regional, national or global good is extremely difficult but also important. Modern power plants often require extensive environmental assessments as part of their construction licensing and approval. Such assessments should take these local effects into account alongside any global impact.

CHAPTER *9*

Life-Cycle Assessment

Life cycle assessment is a method of determining the "cradle-to-grave" impact of a product or installation on the environment. This type of assessment can be applied to a power station to evaluate its environmental impact, and by carrying out similar studies for different types of power plant, alternative technologies can be rated. In the case of a power plant the life cycle assessment will consider all the material, energy, and pollutant flows associated with the construction, operation, and decommissioning of an installation. Provided the analysis is all-encompassing, it will provide a method of directly comparing fossil fuel, nuclear, and renewable generating technologies.

The technique of life cycle analysis (LCA) has been evolving over a number of decades and has recently become the focus for international standards and initiatives such as those put forward by the United Nations Environmental Programme. The International Organization for Standardization (ISO) has now standardized the framework for LCA studies under ISO 14040. Under this standard, a four-part approach to an LCA is the commonly accepted approach today. These four steps, also shown diagrammatically in Fig. 9.1, are as follows:

1. State the purpose of the study and define its boundaries clearly.
2. Quantify the material and energy inputs and all the outputs for each stage of the life cycle of the object under study.
3. Analyze the results in order to determine the life cycle impact on human health and the environment. This is called the life cycle impact assessment (LCIA).
4. Interpret the results of the LCA and the LCIA in order to identify significant issues and areas where improvement might be made.

Even with a standardized approach it can be difficult to compare results from different studies. Establishing, and standardizing, the boundaries of the LCA is especially critical. For example when considering nuclear power, does the study include the energy involved in mining and

Electricity Generation and the Environment. DOI: http://dx.doi.org/10.1016/B978-0-08-101044-0.00009-3

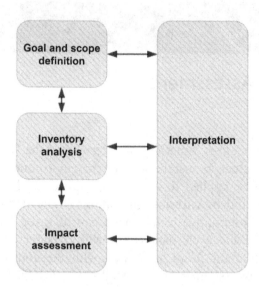

Figure 9.1 The four stages of a life cycle assessment. Wikipedia.

extracting the nuclear fuel or does it only start at the point when the fuel enters the power station? Or, when considering an intermittent renewable energy plant, does the LCA take into account any changes that are made to grid operation to support the intermittent supply?

To make a global comparison of some recent studies possible, the US National Renewable Energy Laboratory (NREL) carried out a review of 2165 published LCAs in 2011 from which it selected 296 as the basis for a systematic study. The results of this review of existing literature were used by the Intergovernmental Panel on Climate Change in a special report on Renewable Energy Sources and Climate Change Mitigation[1]. Some of these results are presented below.

9.1 TYPES OF LIFE CYCLE ANALYSIS

In the case of power plants there are a number of important LCAs that have been identified as useful for evaluating performance. The first of these and often the most salient today is the greenhouse gas impact, otherwise known as the carbon footprint of a power project. A study of this type will seek to quantify the amounts of carbon dioxide

[1]Special Report on Renewable Energy Sources and Climate Change Mitigation, Intergovernmental Panel on Climate Change, 2011.

and other greenhouse gases that are emitted by a generator for each unit of electricity produced. Similar studies can be carried out for other major pollutants such as sulfur, nitrogen oxides, or heavy metals including mercury. Again these will provide figures for the amount of each pollutant released for each unit of electricity produced.

Another useful figure for a power station is the energy payback ratio. A study of this parameter will look at the total amount of energy a power facility produces during its lifetime and compare it to the total amount of energy it consumes. The ratio indicates how many units of power a plant generates for each one it consumes. A closely associated figure is the energy payback time, which shows how long it takes for the power plant to generate as much energy as was (or will be used) during its construction, operation, and decommissioning.

Finally, one further LCA examines the economics of power generation by looking at the lifetime costs of a power plant, of its construction, operation, and decommissioning. This type of analysis is normally called a Levelized Cost of Electricity (LCOE) study and is one of the economic studies underpinning the selection of a power generation technology for a particular situation.

9.2 GREENHOUSE GAS EMISSIONS AND THE CARBON FOOTPRINT

While the focus of the debate on global warming is the combustion of fossil fuels, when looking at power generation, it is important to recognize that all types of power plant have a carbon footprint. This is because, even if they do not actually burn fossil fuel to generate electricity, the materials from which they are made and the vehicles that transport those materials or components to the site all released greenhouse gases.

Table 9.1 presents figures for the life cycle greenhouse gas emissions of all the main power generation technologies. As might be expected, fossil fuel power plants have the largest carbon footprints. Based on the median emission figures that are shown in the table, a coal-fired power station will release 1001 g CO_2 equivalent/kWh. For an oil-fired power station the figure is 840 g CO_2 equivalent/kWh, while a natural gas-fired station will release 469 g CO_2 equivalent/kWh. These figures could be reduced if carbon capture and storage (CCS) technologies were widely available. CCS could reduce the emissions from a

Table 9.1 Life Cycle Greenhouse Gas Emission of Power Generating Technologies[2]	
Technology	Median Life Cycle Greenhouse Gas Emissions (g CO_2equivalent/kWh)
Solar photovoltaic	46
Solar thermal	22
Geothermal	45
Hydropower	4
Ocean	8
Wind	12
Bioenergy	18
Nuclear	16
Natural gas	469
Oil	840
Coal	1001
Source: *IPCC*	

coal plant to 98–396 g CO_2equivalent/kWh, while the natural gas-fired plant would emit 65–245 g CO_2equivalent/kWh with CCS. These CCS figures are not shown in the table.

Compared to these figures, all the renewable technologies in Table 9.1 show much lower emissions. The renewable technology with the largest carbon footprint is solar photovoltaic (solar cell) technology with 46 g CO_2equivalent/kWh. The reason for this relatively high figure is the amount of energy that is required to manufacture pure crystalline silicon for the solar cells. This is an energy-intensive process using electrical power, most of which comes from fossil fuel–fired power stations. If the energy was supplied from renewable sources, this footprint would be much smaller.

Geothermal power also has a relatively high carbon footprint, 45 g CO_2equivalent/kWh. This is a consequence of the carbon dioxide that is released from the geothermal reservoir when the brine is extracted. Solar thermal technology produces life cycle greenhouse gas emissions of 22 g CO_2equivalent/kWh; for a bioenergy plant the emissions are 18 g CO_2equivalent/kWh, for a wind plant the emissions are 12 g CO_2equivalent/kWh, while ocean power technologies are estimated to

[2]The figures are taken from Life Cycle Assessment of Electricity Generation, Eurelectric, 2011, and are based on IPCC figures from the NREL analysis.

release 8 g CO_2equivalent/kWh. The lowest emissions are for a hydropower plant with 4 g CO_2equivalent/kWh.

Part of the reason for the low emissions from hydropower is the long lifetime of a hydropower plant. Because the emissions are mostly associated with its construction, the longer it can produce energy, the lower the carbon footprint will be. Nuclear power plants, which also have long service lives, produce 16 g CO_2equivalent/kWh.

Some LCAs focus solely on greenhouse gas emissions, but it is possible to use an LCA to extract figures for lifetime emissions of pollutants such as sulfur dioxide, nitrogen oxides, and particulates of different sizes. Such studies again show that renewable energy systems produce much lower airborne pollutant levels than fossil fuel plants.

9.3 ENERGY PAYBACK

The energy payback of a power plant is the amount of energy it provides compared to the energy it consumes. This can be the total amount of each over the lifetime of the plant but more usually it is presented as the energy generated for each unit of energy consumed. While the concept is easy to grasp, its significance is less easy to establish. Nevertheless, it provides a yardstick by which to compare different technologies.

If a power plant is going to be an effective source of energy, then it must generate more energy than it consumes over its lifetime. When calculating the energy payback ratio, the energy contained in the energy source that the technology exploits—coal, oil, or gas in the case of fossil fuels, wind, sunlight, or flowing water in the case of renewable sources—is not usually included in the calculations. If these are included, then fossil fuel power plants will show a higher energy payback ratio than renewable energy plants because the former are generally more efficient at capturing energy from the source.

As with carbon footprint calculations, the lifetime assumed for a power plant is important when considering the energy payback. The longer a power plant operates, the more energy it will produce relative to the energy it consumes and the better its energy payback will be.

Table 9.2 contains figures for the energy payback ratio of the main types of power generating plant when the energy source is *not* included

Table 9.2 Energy Payback Ratio[3]		
Technology	Minimum Energy Payback Ratio	Maximum Energy Payback Ratio
Supercritical coal	2.9	10.1
Natural gas combined cycle	2.5	8.6
Light water reactor	1.5	16.0
Solar photovoltaic	0.8	47.4
Solar thermal	1.0	10.3
Geothermal	2.5	14.0
Wind power	5.0	40.0
Hydropower	6.0	280.0
Source: *IPCC*		

in the calculation. The table contains both minimum and maximum values. The minimum values all fall between 0.8 and 6.0 and represent worst case or poorly performing plants. The maximum values are more indicative of the performance of good examples of each technology.

The best performing plant is a hydropower plant with an energy payback ratio of 280.0. Good hydropower plants will have very long lifetime (in this case the lifetime is 70 years on average), which makes them extremely effective sources of energy. Solar photovoltaic power plants, with a maximum energy payback ratio of 47.4, and wind power, with an energy payback ratio of 40.0, also perform well.

Compared to these three, the others in the table are less impressive. For a light water nuclear power plant the maximum energy payback ratio is 16, and for a geothermal plant it is 14. Solar thermal power plants have an energy payback ratio of 10.3, slightly higher than that of a supercritical coal-fired power station. The natural gas-fired combined cycle power plant is at the bottom, with an energy payback ratio of 8.6.

Closely related to the energy payback ratio is the energy payback time. Some energy payback times are shown in Table 9.3. Again the table contains minimum and maximum figures. In this case the lower the figure, the better the performance. The column containing the maximum energy payback times shows that solar plants have potentially the

[3]The figures are taken from Life Cycle Assessment of Electricity Generation, Eurelectric, 2011, and are based on IPCC figures from the NREL analysis.

Table 9.3 Energy Payback Time[4]		
Technology	Minimum Energy Payback Time (years)	Maximum Energy Payback Time (years)
Supercritical coal	1.0	2.6
Natural gas combined cycle	1.2	3.6
Light water reactor	0.8	3.0
Solar photovoltaic	0.2	8.0
Solar thermal	0.7	7.5
Geothermal	0.6	3.6
Wind power	0.1	1.5
Hydropower	0.1	3.5
Source: IPCC		

worst performance with maximum energy payback times of 7–8 years. However, the minimum payback times are around 6 months or less.

Excepting these two, the maximum energy payback times in the table are all under 3.6 years while the minimum times are all under 1.2 years, the latter for a natural gas-fired combined cycle power plant. A plant that had an energy payback time that was longer than its service life would clearly be uneconomical in energy terms. While some of the worst-case energy payback ratios represent a large fraction of a plant lifetime, the minimum values are all acceptable in energy payback terms.

9.4 LEVELIZED COST OF ELECTRICITY

The cost of electricity from a power plant of any type depends on a range of factors. First, there is the cost of building the power station and buying all the components needed for its construction. In addition, most large power projects today are financed using loans, so there will also be a cost associated with paying back the loan, with interest. Then, there is the cost of operating and maintaining the plant over its lifetime, including fuel costs if the plant burns a fuel. Finally, the overall cost equation should include the cost of decommissioning the power station once it is removed from service.

[4]The figures are taken from Life Cycle Assessment of Electricity Generation, Eurelectric, 2011, and are based on IPCC figures from the NREL analysis.

It would be possible to add up all these cost elements to provide a total cost of building and running the power station over its lifetime, including the cost of decommissioning, and then dividing this total by the total number of units of electricity that the power station actually produced over its lifetime. The result would be the real lifetime cost of electricity from the plant. Unfortunately such a calculation could only be completed once the power station was no longer in service. From a practical point of view, this would not be of much use. The point in time at which the cost-of-electricity calculation of this type is most needed is before the power station is built. This is when a decision is made to build a particular type of power plant, based normally on the technology that will offer the least cost electricity over its lifetime.

To get around this problem, economists have devised a model that provides an estimate of the lifetime cost of electricity before the station is built. Of course, because the plant does not yet exist, the model requires that a large number of assumptions be made. To make this model as useful as possible, all future costs are also converted to the equivalent cost today by using a parameter known as the discount rate. The discount rate is almost the same as the interest rate and relates to the way in which the value of one unit of currency falls (most usually, but it could rise) in the future. This allows, for example, the cost of replacement of a plant component 20 years into the future to be converted into an equivalent cost today. The discount rate can also be applied the cost of electricity from the power plant in 20 years' time.

The economic model is called the LCOE model. It contains a lot of assumptions and flaws but is the most commonly used method available for estimating the cost of electricity from a new power plant. One particular problem is that the model does not take into account cost risks. For example, the cost of natural gas can fluctuate widely, so that it may be cheap to buy gas when a plant is built but 5 years later the cost is too high that operation of the plant is uneconomical. The level at which the discount rate is set can also be problematical. It is typical to use a discount rate of 5% and 10% in calculations. However, in the middle of the second decade of the 21st century the actual interest rate is close to zero.

It can also be difficult to compare the LCOE of different technologies. In an attempt to offer additional guidance, the US Energy Information Administration has recently also started to calculate the

Levelized Avoided Cost of Electricity. This gauges the cost to the grid to supply the electricity if a new power plant of a particular type was not built. The levelized avoided cost provides a guide to the economic value of a power plant under the specific circumstances under which it is being considered.

Levelized costs are included in all the technology-specific books in this series, and comparative values will not be included here. There are a variety of sources for comparative figures including the US Energy Information Administration that provides figures annually and the International Energy Agency that assesses the global costs and variations in LCOE in a report it publishes every 5 years.

CHAPTER 10

Externalities: Putting a Price on Environmental Damage

Life cycle studies can quantify the environmental impact of different technologies in terms of the amount of carbon dioxide released or the amount of energy that a plant produces for each unit it consumes. However, these figures do not, in themselves, provide any idea of the external cost of these emissions to the world at large. The external social and environmental cost of power generation, usually called their externalities, will be examined in this chapter.

Externality is a term derived from economics where it refers to the unpriced costs of the side effects of a production process. These costs are usually imposed on third parties rather than being underwritten by the owner of the process. They include factors such as impact on climate, human health, crops and buildings, and biodiversity. The range of externalities associated with power generation is enormous. For example, it should encompass the effects of the prospecting, mining, and transportation of fuel; the effect of any emissions from a power plant; and the impact of waste from the plant. For a renewable power plant it must take account of the effect of the electricity from the plant on the grid into which it is supplied. If the renewable plant requires fossil fuel−fired backup, the externalities associated with that should be accounted for too.

Given this range, and the difficulty in many instances of estimating the overall external cost of each component, the values put on externalities have ranged widely. To create a more objective framework, an international project called ExternE was launched in 1991 by the European Commission and the US Department of Energy. The latter withdrew in 1994 but the work of ExternE has continued, now primarily through a project called New Energy Externalities Development for Sustainability (NEEDS).

Electricity Generation and the Environment. DOI: http://dx.doi.org/10.1016/B978-0-08-101044-0.00010-X

The objective of ExternE and NEEDS is to put a monetary value on the external impact of power generation. This represents a cost that the producers of the electricity do not pay and which is therefore borne by the society as a whole. It is then up to policy makers and governments to find ways of introducing methods by which the producer of the external effect pays for its impact. This has been referred to as "the polluter pays." So far the efforts in this direction have been limited.

10.1 PUTTING A PRICE ON EXTERNALITIES

One of the most difficult tasks for any externalities study is to put a price on an external impact. How, at one extreme, do you value the loss of a life? Or at the other extreme, do you put a cost on low-level respiratory effects of airborne pollution?

Typical valuations of this type, from ExternE and based on the price—year 2000, are €1,000,000 for a prevented fatality, €50,000 for each year of life lost, €190,000 for a case of chronic bronchitis, and €670 for each emergency hospital visit as a result of respiratory illness[1]. Such figures may seem arbitrary but they are based on structured analysis. However, the range and complexity of health effects makes this a particularly difficult area.

Another way of calculating the cost on external impacts is to place a value on each ton of a particular pollutant emitted into the environment. This also requires a complex calculation to arrive at a realistic cost. In ExternE the greenhouse damage cost of the emission of 1 t of carbon dioxide was put at €19. For other pollutants the cost is much higher. For example, the cost of damage to health arising from the emission of 1 t of particulate material is put at €25,000–€75,000. For sulfur dioxide the cost is €4500–€16,000/t, while for nitrogen oxides it is €4200–€11,000/t. Then there are effects on plants, other animals, and on buildings. In the middle of the first decade of the 21st century the estimated aggregate annual cost of externalities resulting from power generation across the European Union (EU) was €80 billion.

---

[1]The figures in this chapter are based on date from the ExternE project, published in The Hidden Costs of Electricity: Externalities of Power Generation in Australia, Australian Academy of Technological Sciences and Engineering, 2009.

10.2 THE EXTERNAL COST OF POWER GENERATION

To put these externalities into perspective for different power generating technologies, Table 10.1 contains figures that show the range of external costs for a range different technologies in the European Union, as estimated by ExternE. These figures, which are based on studies carried out from the start to the middle of the first decade of the 21st century, can be compared to the cost of electricity from similar power stations in the same period to give an idea of their scale. The levelized cost of electricity at the time these figures were current would have been around €30—€40/MWh for power from a fossil fuel—fired power plant[2]. Wind power would have cost more than this in the early part of this century but has fallen dramatically since.

It will be immediately clear from the table that the largest external costs from ExternE are all associated with fossil fuel combustion. Of those included in the table, oil with a cost range of €30—€110/MWh and coal with €20—€100/MWh are the highest. Peat at €20—€50/MWh and natural gas (€10—€40/MWh) are close behind. As has been made clear in earlier chapters of this book, these fuels are dirty and they produce significant quantities of carbon dioxide. If these external costs were to be internalized and the generators were required to pay them,

Table 10.1 External Costs in the European Union From the ExternE Project[3]	
Energy Source	**External Cost (€/MWh)**
Coal and lignite	20—100
Peat	20—50
Oil	30—110
Natural gas	10—40
Nuclear	2—7
Biomass	0—50
Hydropower	1—10
Solar photovoltaic	6
Wind	1—3
Source: *EU*	

[2]The Future of Power Generation, Paul Breeze, Business Insights, 2005.
[3]The figures in the table are from the ExternE project, published in The Hidden Costs of Electricity: Externalities of Power Generation in Australia, Australian Academy of Technological Sciences and Engineering, 2009.

then electricity from these power plants would cost two to three times the amount it costs today.

Nuclear power has a much lower external cost, according to ExternE, with an estimated figure of between €2 and €7/MWh. However, nuclear power presents some additional difficulties because of the potential risks associated with the technology, which are not generally costs in an externalities study.

Aside from nuclear power, renewable technologies generally have the lowest external costs. Hydropower has an external cost range of €1–€10/MWh, while for wind power the range is €1–€3/MWh. The external cost of solar photovoltaic, from a single country—Germany— was estimated to be €6/MWh. Biomass combustion, which can involve the release of atmospheric pollutants during operation, has a wide range of external costs, from 0 to €50/MWh.

10.3 INTERNALIZING EXTERNAL COSTS

In theory it is possible to internalize the external costs associated with different power generation technologies. In practice it has so far proved difficult to do so.

Because the damage caused by the power plants is inflicted on society as a whole, it is obvious that the society must be paying the price. The most logical means of correcting this would be to impose a tax on the generation of power based on the type of power station that is generating it and using the income to deal with the results of the emissions.

There are a number of problems with this simple approach. How is the income to be apportioned and how will it be directed to the right targets? Is part of the tax used to help fund a health service or is it put into a fund for individuals who are assessed to have been affected? The same applies, but more broadly, to the damage to the environment, to crops and plants, to biodiversity and to buildings. In addition, much of the damage being caused by greenhouse gases is global rather than local. How can that be taken into account when taxes are raised at a national level?

Then there is the difficulty of raising the price of electricity. Most governments are reluctant to take measures that will cause major popular upset, and increasing the cost of electricity is likely to be a measure with the effect. In addition, more candidly, governments cannot be trusted to use the money they collect in this way to alleviate the problem for which it is intended. They are just as likely to use it to build roads or provide tax cuts.

Rather than taxing the output of power stations, another approach would be to tax fuel for power stations. A carbon tax of this sort has been applied in some countries. However, the main objections already outlined above apply to this approach too. In practice the main value of costing externalities today is in highlighting that cost and spurring action to prevent the external consequences, such as power plant emissions, which cause the damage.

CHAPTER 11

The Bottom Line

The generation of electric power is one of the most important industrial activities taking place across the globe today. Modern life cannot function without electricity and it has helped to raise the living standards and the educational standards of most people on earth. At the same time the production of electricity has a profound effect on the global environment. Balancing the need for electricity with its adverse effects has become a major challenge. How that challenge is met will determine the future shape of life on our planet.

The main challenge is to control global warming. That will require drastic reductions in the amount of carbon dioxide that is pumped into the atmosphere. Reduction can be achieved by shifting from fossil fuel power generation to the production of electricity based on renewable energy sources such as wind, sun, and water. There is also an alternative, exploiting technique that is capable of capturing the carbon dioxide from fossil fuel power stations and sequestering it inside the earth. In principle both can renewable generation and the use of carbon capture with fossil fuel combustion can help. However, in practice these latter technologies are not advancing quickly enough and are likely to be left behind. The momentum within the industry appears to be shifting towards renewable generation.

Meanwhile, renewable technologies have already proved that they are capable of providing the power the world needs, but they cannot do so overnight. To convert the globe entirely to renewable energy is the work of several generations. In addition, there are still tactical issues to be resolved if renewable power is to provide the entire world's electricity. This means that fossil fuels will continue to be burned for power generation for decades to come, with or without carbon capture.

There are economic issues to be dealt with too. The electricity sector in many countries of the world is based on market principles where decisions are made primarily based on cost. Renewable energy is not

Electricity Generation and the Environment. DOI: http://dx.doi.org/10.1016/B978-0-08-101044-0.00011-1

always the cheapest source available for new generation, so it requires government policy and intervention to steer generation towards renewable energy. This represents one difficulty. Another is related to vested interests and how they are to be handled. The fossil fuel industry is a massive global industry, and it employs millions of people. If fossil fuels are phased out, then so too is the employment and wealth creation associated with them.

Phasing out fossil fuels will eliminate most of the major sources of externalities associated with the power generation industry today. However, the growth of renewable generation may lead externalities that are associated with the new technologies growing in importance.

Managing the transition from fossil fuels to renewable energy at the technological level, at the economic level, and at the social level will also be the work of generations. The changes have only just started but if, and as they become embedded they are likely to have a profound effect on the whole electricity industry. It seems clear from the perspective of the beginning of the 21st century that these changes are already underway. Their impact will be profound. What the industry will look like when the process has ended will not be visible for several decades to come.

Note: Page numbers followed by "*f*" and "*t*" refer to figures and tables, respectively.

Printed in the United States
By Bookmasters